ENERGY POLICIES, POLITICS AND PRICES

U.S. PETROLEUM REFINING

PETCOKE ISSUES, SMALL REFINERY OPPORTUNITIES AND INDUSTRY BACKGROUND

ENERGY POLICIES, POLITICS AND PRICES

Additional books in this series can be found on Nova's website under the Series tab.

Additional e-books in this series can be found on Nova's website under the e-book tab.

ENERGY POLICIES, POLITICS AND PRICES

U.S. PETROLEUM REFINING

PETCOKE ISSUES, SMALL REFINERY OPPORTUNITIES AND INDUSTRY BACKGROUND

ROGER M. WATERS
EDITOR

nova publishers
New York

Library of Congress Cataloging-in-Publication Data

ISBN: 978-1-63463-336-9

Published by Nova Science Publishers, Inc. † New York

CONTENTS

PREFACE

Some Members of Congress have expressed concern about storage and management of petroleum coke (petcoke) in their districts. Despite a lack of data on observed health impacts, local concerns have included potential human health and environmental impacts of fugitive dusts and runoff into waterways. Broader concerns have also been raised about the carbon dioxide (greenhouse gas) emissions from petcoke combustion. This book discusses the petcoke issues, small refinery opportunities and provides a background on the industry.

Chapter 1 – In early 2013, media outlets around Detroit, Michigan began publishing stories about large piles of petroleum coke stored along the Detroit Riverfront. Petroleum coke (petcoke) is a black-colored solid composed primarily of carbon, and may contain limited amounts of elemental forms of sulfur, metals and non-volatile inorganic compounds. Petcoke is essentially chemically inert. Petcoke exposure is considered to pose few human health or environmental risks, but may present significant nuisance concerns. The material in Detroit was the byproduct of the nearby Marathon Refinery's processing of heavy crude oils derived, in part, from Canadian oil sands deposits. The situation gained national attention with the publication of an article in the *New York Times* ("A Black Mound of Canadian Oil Waste Is Rising over Detroit," *New York Times,* May 17, 2013). The piles of petcoke sparked local concerns over the potential impacts of the material on human health and the environment, and whether these concerns were adequately addressed by local, state, and federal regulations. As petroleum refining is a nationwide commercial industry, these concerns may arise in other regions.

Petcoke is a co-product of several distillation processes used in refining heavy crude oil. Nearly half of U.S. petroleum refineries (56 or more) use a coking process to convert heavy crude oils into refined petroleum products, and more refineries may follow suit to take advantage of the supply of heavy crude oils from Canada's oil sands projects. Although it is a refining co-product, petcoke has economic value as both a heating fuel and raw material in manufacturing. In 2012, the U.S. Energy Information Administration reported that U.S. refineries produced in excess of 56 million metric tons of petcoke, of which 80% was exported.

The U.S. Environmental Protection Agency has surveyed the potential human health and environmental impacts of petcoke through its High Production Volume (HPV) Challenge Program and found the material to be highly stable and non-reactive at ambient environmental conditions. Most toxicity analyses of petcoke find it has a low potential to cause adverse effects on aquatic or terrestrial environments as well as a low health hazard potential in humans, with no observed carcinogenic, reproductive, or developmental effects. Cases of repeated-dose and chronic inhalation of fugitive dust (as generated during petcoke handling and storage) in animal studies do appear associated with respiratory inflammation. Emissions from the combustion of petcoke, however, can have impacts on human health and the environment, including the release of common pollutants, hazardous substances, and high levels of the greenhouse gas, carbon dioxide.

While some federal statutes address certain environmental impacts of petcoke's life-cycle, most regulatory action and oversight has been undertaken at the state and local levels, generally through facility-specific permitting requirements. Federally, petcoke is exempted from classification as either a solid or hazardous waste under the Resource Conservation and Recovery Act (RCRA) and is not considered a hazardous substance under the Comprehensive Environmental Response, Compensation, and Liability Act (CERCLA). Petcoke facilities may be regulated under certain provisions of the National Pollutant Discharge Elimination System (NPDES) permit program, as authorized by the Clean Water Act (CWA), if it is determined that runoff from sites where it is stored has the potential to transport the substance to nearby surface waters. The handling of petcoke may also create instances of reduced air quality due to releases of fugitive dust into the atmosphere. Most of the impacts of fugitive dust are localized; and thus, much of the regulatory oversight is implemented at the local and state level. Whether such oversight is providing adequate protection is among the issues that have been raised.

Chapter 2 – The last refinery constructed in the United States opened in 1977. Since the mid-1980s, some 150 have closed as part of an industry-wide consolidation. Over the same time, the remaining refineries expanded their operational capacity by 23% to keep up with increasing demand. Current U.S.

refining capacity appears to satisfy if not exceed demand as the increasing export of refined petroleum products would seem to suggest. Notwithstanding the current surplus capacity, opportunities for new refineries appear to have emerged as the result of the rise in production of U.S. light-sweet crude oil from unconventional resources such as North Dakota's Bakken and Texas' Eagle Ford formations. These new resources have revitalized some refinery operations that formerly depended on imported light crude oil, thereby making the smaller refineries more competitive with large refineries that process more widely available heavy-sour crude oil. Rising domestic crude oil production has not only led U.S. refineries to export their petroleum products, but has led some oil producers to attempt to bypass refining and export crude oil condensates directly. However, Department of Commerce regulations currently restrict crude oil exports. Whether condensates fall under a refined product classification or crude oil remains an unresolved issue, and one that Congress may choose to take up. Congress may also consider whether small businesses face inherent disadvantages in entering an industry dominated by large complex refinery operators.

While some arguments remain for new refinery construction, some argue that economic and regulatory barriers hinder new construction. Small refining enterprises may suffer certain economic disadvantages in an industry driven by the economy of scale. Furthermore, opportunities may depend on geographic location, regional markets, and source of crude oil. The barriers from regulatory requirements remain less clear. Export restrictions certainly affect potential product markets that could support new refining capacity. Whether these markets would spur new refinery construction is a separate question.

The history of refinery expansions suggests that environmental requirements do not generally pose a significant barrier to refinery construction, but local challenges vary (e.g., locations with currently unhealthy air quality). Small refineries face many of the same economic, market, and environmental factors that affect large refineries, but they may also benefit from exemptions in complying with certain federal regulations. As with any industrial facility involving potential air emissions, water use, and waste generation, certain environmental permitting and regulatory requirements apply to the construction and operation of refineries. The specific requirements, and the degree of difficulty or ease in meeting them, will depend not only on the nature of the project, but perhaps more importantly on the location of the project.

Small business opportunities in refining may strongly depend on a number of factors, including regional demand for refined products, available crude oil supplies, and unique regional environmental regulations, among others. A "small business" in the oil refining business is defined differently in different statutes. In addition, some fuels are easier to produce than others are. While evaluating opportunities is beyond the scope of this report, some key opportunities have recently emerged. Perhaps the most discussed potential opportunity for constructing a small topping or hydroskimming refinery is in the mid-continent region (North Dakota, Colorado) to process these new sources of light sweet crude. Other opportunities may lie in installing condensate splitter and stabilizer plants to handle North Dakota and Texas expanding production of light sweet crude oil and condensate.

Chapter 3 – A decade ago, 158 refineries operated in the United States and its territories and sporadic refinery outages led many policy makers to advocate new refinery construction. Fears that crude oil production was in decline also led to policies promoting alternative fuels and increased vehicle fuel efficiency. Since the summer 2008 peak in crude oil prices, however, the U.S. demand for refined petroleum products has declined, and the outlook for the petroleum refining industry in the United States has changed.

In response to weak demand for gasoline and other refined products, refinery operators have begun cutting back capacity, idling, and, in a few cases, permanently closing their refineries. By current count, 124 refineries now produce fuel in addition to 13 refineries that produce lubricating oils and asphalt. Even as the number of refineries has decreased, operable refining capacity has actually increased over the past decade, from 16.5 million barrels/day to over 18 million barrels/day. Cyclical economic factors aside, U.S. refiners now face the potential of long-term decreased demand for their products. This is the result of legislative and regulatory efforts that were originally intended, in part, to accommodate the growing demand for petroleum products, but which may now displace some of that demand. These efforts include such policies as increasing the volume of ethanol in the gasoline supply, improving vehicle fuel efficiency, and encouraging the purchase of vehicles powered by natural gas or electricity.

Since the Clean Air Act Amendments, 15 distinctly formulated boutique fuels are required in portions of 12 states. H.R. 392, the Boutique Fuel Reduction Act of 2009, would further amend` the Clean Air Act to add temporary waivers for boutique fuels due to unexpected problems with distribution and give EPA authority to reduce the number of boutique fuels. The 2005 Energy Policy Act created the Renewable Fuel Program to substitute increasing volumes of renewable fuel for gasoline. The 2007 Energy Independence and Security Act expanded the program to cover transportation fuels in general, extended the program to calendar year 2022, and increased the target volume to 36 billion gallons renewable fuel annually. The 2008 Food, Conservation and Energy Act of 2008 reduced some of the federal subsidies and tax breaks favoring ethanol production. A 2007U.S. Supreme Court ruling found that EPA has the authority under the Clean Air Act to regulate carbon dioxide (CO2) emissions from automobiles. Though the ruling applied to automobiles, it had wider implications. In response to the FY2008 Consolidated Appropriations Act (H.R. 2764; P.L. 110-161), EPA issued the Mandatory Reporting of Greenhouse Gases Rule that requires suppliers of fossil fuels or industrial greenhouse gases (GHG), manufacturers of vehicles and engines, and facilities that emit 25,000 metric tons or more per year of GHG emissions to submit annual reports to EPA. H.R. 2454, The American Clean Energy and Security Act of 2009 (passed in the House June 26, 2009) would amend the Clean Air Act by establishing a "cap-and-trade" system designed to reduce greenhouse gas emissions (GHG) and would cap emissions from refineries and allow trading of emissions permits ("allowances"). As proposed, H.R. 2454 would require U.S. refiners to purchase emission credits for both their stationary emissions and the subsequent combustion of their fuels (predominantly consumed in the transportation sector). S. 3663, introduced in August 2010, would establish a Natural Gas Vehicle and Infrastructure Development Program to promote natural gas as an alternative transportation fuel in order to reduce domestic oil use.

The prospect of declining motor-fuel demand may persuade operators to idle, consolidate, or permanently close refineries.

In: U.S. Petroleum Refining
Editor: Roger M. Waters
ISBN: 978-1-63463-336-9

Chapter 1

PETROLEUM COKE: INDUSTRY AND ENVIRONMENTAL ISSUES*

Anthony Andrews and Richard K. Lattanzio

SUMMARY

In early 2013, media outlets around Detroit, Michigan began publishing stories about large piles of petroleum coke stored along the Detroit Riverfront. Petroleum coke (petcoke) is a black-colored solid composed primarily of carbon, and may contain limited amounts of elemental forms of sulfur, metals and non-volatile inorganic compounds. Petcoke is essentially chemically inert. Petcoke exposure is considered to pose few human health or environmental risks, but may present significant nuisance concerns. The material in Detroit was the byproduct of the nearby Marathon Refinery's processing of heavy crude oils derived, in part, from Canadian oil sands deposits. The situation gained national attention with the publication of an article in the *New York Times* ("A Black Mound of Canadian Oil Waste Is Rising over Detroit," *New York Times,* May 17, 2013). The piles of petcoke sparked local concerns over the potential impacts of the material on human health and the environment, and whether these concerns were adequately addressed by local, state, and federal regulations. As petroleum refining is a nationwide commercial industry, these concerns may arise in other regions.

* This is an edited, reformatted and augmented version of a Congressional Research Service publication, No. R43263, dated October 29, 2013.

Petcoke is a co-product of several distillation processes used in refining heavy crude oil. Nearly half of U.S. petroleum refineries (56 or more) use a coking process to convert heavy crude oils into refined petroleum products, and more refineries may follow suit to take advantage of the supply of heavy crude oils from Canada's oil sands projects. Although it is a refining co-product, petcoke has economic value as both a heating fuel and raw material in manufacturing. In 2012, the U.S. Energy Information Administration reported that U.S. refineries produced in excess of 56 million metric tons of petcoke, of which 80% was exported.

The U.S. Environmental Protection Agency has surveyed the potential human health and environmental impacts of petcoke through its High Production Volume (HPV) Challenge Program and found the material to be highly stable and non-reactive at ambient environmental conditions. Most toxicity analyses of petcoke find it has a low potential to cause adverse effects on aquatic or terrestrial environments as well as a low health hazard potential in humans, with no observed carcinogenic, reproductive, or developmental effects. Cases of repeated-dose and chronic inhalation of fugitive dust (as generated during petcoke handling and storage) in animal studies do appear associated with respiratory inflammation. Emissions from the combustion of petcoke, however, can have impacts on human health and the environment, including the release of common pollutants, hazardous substances, and high levels of the greenhouse gas, carbon dioxide.

While some federal statutes address certain environmental impacts of petcoke's life-cycle, most regulatory action and oversight has been undertaken at the state and local levels, generally through facility-specific permitting requirements. Federally, petcoke is exempted from classification as either a solid or hazardous waste under the Resource Conservation and Recovery Act (RCRA) and is not considered a hazardous substance under the Comprehensive Environmental Response, Compensation, and Liability Act (CERCLA). Petcoke facilities may be regulated under certain provisions of the National Pollutant Discharge Elimination System (NPDES) permit program, as authorized by the Clean Water Act (CWA), if it is determined that runoff from sites where it is stored has the potential to transport the substance to nearby surface waters. The handling of petcoke may also create instances of reduced air quality due to releases of fugitive dust into the atmosphere. Most of the impacts of fugitive dust are localized; and thus, much of the regulatory oversight is implemented at the local and state level. Whether such oversight is providing adequate protection is among the issues that have been raised.

BACKGROUND

Some Members of Congress have expressed concern about storage and management of petroleum coke (petcoke) in their districts. Despite a lack of data on observed health impacts, local concerns have included potential human health and environmental impacts of fugitive dusts and runoff into waterways. Broader concerns have also been raised about the carbon dioxide (greenhouse gas) emissions from petcoke combustion.

Petcoke is the co-product of several processes used in petroleum refining to upgrade "residuum" into gasoline and middle distillate-range fuels. Residuum (or resid) remains after refineries initially distill heavy crude oils. Petcoke is a black-colored solid composed primarily of carbon, and may contain limited amounts of elemental forms of sulfur, metals, and non-volatile inorganic compounds.

The petroleum industry and federal regulators characterize petcoke as a "co-product" because it may have some commercial value as a boiler fuel and as a raw material in manufacturing. Nearly half of U.S. petroleum refineries employ "coking" processes. Refineries also produce petcoke as a by-product of catalysis, which refineries later consume as a fuel.

In addition to the existing suite of coking refineries, other refineries may add coking processes to take advantage of increased supplies of heavy crude oils from Canada's oil sand projects. Meanwhile, newly available light crude oil from U.S. unconventional shale projects and the Texas Permian Basin is leading some coking refineries to cut back on coking. At present, it is uncertain whether petcoke production will increase, remain steady or even decline, given the changing slate of U.S. crude oil supplies. Nevertheless, the export and demand for U.S. petcoke has been rising recently.

Community stakeholders have grown concerned over the potential effects on public health and the environment related to the production, storage, transportation, and use of petcoke. Some of these impacts include concerns over air quality due to fugitive dust, water quality due to run-off, and the potential for toxic and other emissions (including greenhouse gas emissions) from its combustion as a fuel source. In light of these concerns, industry, regulators, and compliance officers may be interested in best practices related to the storing, containing, and managing of petcoke.

Petcoke Uses

Petcoke may be combusted as fuel in industrial and power generating plants. Cement plants and power plants are currently the two greatest consumers of petcoke. There is some limited use as space heating and in commercial brick kilns in Europe, and a small but emerging market as a metallurgical coal blending component for the steel industry. In the United States, the high sulfur content may limit the petcoke in a coal/petcoke blend in a plant designed for coal. However, more recently designed Circulating Fluidized Bed (CFB) boilers can accommodate 100% high sulfur coke.[1]

Fuel grade petcoke can substitute for "steam coal" in power plant boilers, having the advantage of a higher heating value (discussed below). Conventional coal-fired boilers can blend petcoke with steam coal, and newer boiler designs have replaced steam coal with petcoke entirely.[2] Cement plants consume fuel-grade petcoke in rotary kilns.

Anode grade calcined petcoke is the principal raw material used in manufacturing carbon anodes for use in aluminum smelting. The anodes act as conductors of electricity and as a source of carbon in the electrolytic cell that reduces alumina into aluminum metal. Carbon anode manufacturers, predominantly captive operations of aluminum smelting companies, purchase anode grade calcined petcoke, mix it with pitch binders, press the mixture into blocks, and then bake the mixture to form a finished, hardened carbon anode.

Petcoke Composition

Petcoke is composed primarily of carbon. The specific chemical composition of petcoke depends on the composition of the petroleum feedstock used in refining. Petcoke impurities (i.e., the non-elemental carbonaceous substances) include some residual hydrocarbons left over from processing (referred to as volatiles), as well as elemental forms of nitrogen, sulfur, nickel, vanadium, and other heavy metals. These impurities exist as a hardened residuum captured within coke's carbon matrix. **Table 1** provides an observed range of petcoke properties for green and calcined petcoke.

Table 1. Petcoke Elemental Composition By Weight Percent

Composition	Green	Calcined
Carbon	89.58–91.80	98.40
Hydrogen	3.71–5.04	0.14
Oxygen	1.30–2.14	0.02
Nitrogen	0.95–1.20	0.22
Sulfur	1.29–3.42	1.20
Ash (including heavy metals such as nickel and vanadium)	0.19–0.35	0.35
Carbon-Hydrogen Ratio	18:1–24:1	910:1

Source: U.S. Environmental Protection Agency, Screening-Level Hazard Characterization, Petroleum Coke Category, June 2011; and H.W. Nelson, *Petroleum Coke Handling Problems*, 1970.

Notes: The process of "calcining" converts green coke to almost pure carbon, with a defined structure to produce carbon anodes for the aluminum industry.

Petcoke Compared to Metallurgical Coke and Coal

Petcoke has a significantly high heating value compared to metallurgical coke (metcoke) and bituminous coals (see **Table 2**). The higher heating value comes at the cost of higher sulfur and nitrogen content, however. Ash content is relatively low, compared to coal, but much of it is in the form of heavy metals. Due to the severe thermal environment in which petcoke forms, there is very little combustible volatile material. The low volatile content, in comparison to coal and other fossil fuels, makes petcoke more difficult to ignite and sustain combustion.[3]

Bituminous coal includes two subtypes: thermal and metallurgical.[4] Metallurgical coke is made from low ash, low sulfur bituminous coal, with special coking properties. To produce metcoke, special coke ovens heat metallurgical grade coal at temperatures of 1,000°F to 2,000°F to fuse fixed carbon and inherent ash, and drive off most of the volatile matter.[5] Approximately 1.5 tons of metallurgical coal will produce one ton of metcoke. The final product is a nearly pure carbon source with sizes ranging from basketballs (foundry coke) to a fine powder (coke breeze).

Table 2. Petcoke vs. Metcoke and Coal Heating Value and Price

Fuel	Coal Rank	Btu / lb.	$/Short Ton
Petcoke	n.a.	14,200	See Note
Metcoke	Metallurgic	12,600	171.51
Steam Coal			
Pittsburgh #8	Bituminous	13,000	68.25
Illinois #6	Bituminous	11,000	45.40
Powder River Basin	Sub-bituminous	8,800	10.30

Source: MIT, The Future of Coal Appendices, P5, http://web.mit.edu/coal/The_Future_of_Coal_Appendices.pdf, and EIA, Coal News and markets, http://www.eia.gov/coal/news_markets/.

Notes: Steam Coal Prices as of July 19, 2013. Petcoke prices track steam coal prices, but at a discount that may range from 15% to 85%. Recent prices have been closer to 67% of steam coal prices.

Petcoke Grades

The coking processes described above produce "green coke," which then requires additional thermal processing to remove any residual hydrocarbons (volatile matter) to increase the percentage of elemental carbon. Thermal processing lowers the potential toxicity of the coke. Depending on the coking operation temperatures, length of coking-time, and quality of the crude oil feedstock, one of several grades of petcoke can be produced:

- *sponge coke*, the most common type of regular-grade petcoke, used as a solid fuel (see **Figure 1**);
- *needle coke,* a premium-grade coke made from special petroleum feedstock, used in the manufacture of high-quality graphite electrodes for the steel industry;
- *shot coke,* produced from heavy petroleum feedstock, used as fuel, but less desirable than sponge coke (see **Figure 2**);
- *purge coke*, produced by flexi-coking, used as a fuel in coke-burning boilers; or
- *catalyst coke,* carbon deposited on catalysts, used in various refining processes and burned off and used as a fuel in the refining process; not recoverable in a concentrated form.

Source: John D. Elliott, Shot Coke: Design & Operations, http://www.fwc.com/publications/tech_papers/oil_gas/shotcoke.pdf.

Figure 1. Sponge Coke.

Source: John D. Elliott, Shot Coke: Design & Operations, http://www.fwc.com/publications/tech_papers/oil_gas/shotcoke.pdf.

Figure 2. Shot Coke; Partially Crushed.

COKING REFINERIES AND OUTPUTS

The fleet of petroleum refineries operating throughout the United States has steadily declined in the past several decades as refining capacity has become concentrated in larger refineries. At present, some 115 refineries (and refinery complexes) produce over 17 million barrels per day of motor fuels and other petroleum products. Nearly half (56) have the coking capacity to convert heavy crude oils[6] (see the **Appendix** to this report).

Coking capacity has been concentrated in refineries operating along the Gulf Coast, the historic refining center of the United States and primary destination for crude oil imports. However, to take advantage of the increasing supply of heavier crude oils from Canada's oil sands projects, several Midwest refineries have added coking conversion capacity.

U.S. refineries have the capacity to process 2.5 million barrels per day of petroleum resid. The Gulf Coast not only represents the greatest refining capacity (9.3 million barrels per day), it also represents also the greatest coking capacity (1.5 million barrels per day).

U.S. petcoke production has remained constant over the last decade for the reason that refining capacity has remained constant (see **Figure 3**). In 2012, the U.S. Energy Information Administration (EIA) reported that U.S. refineries produced 42 million metric tons of marketable petcoke and another 15 million metric tons of catalyst coke.[7] For the purpose of comparison, the United States produced 9.3 million tons of coke from metallurgical grade coal[8] and more than 1.2 billion tons of coal in 2012.[9]

Overall, petcoke production reflects refinery capacity utilization rate, which represents the use of the refinery atmospheric crude oil distillation units. The rate is calculated by dividing the gross input to these units by the operable refining capacity of the units. The utilization rate has averaged from 82% to 88%.[10] U.S. refineries have been producing approximately 40 million metric tons of marketable petcoke annually over the period of 2007 through 2012. Refineries, however, need enough light-heavy price spread (LHS) between coker feedstock (heavy resid) and light products (gasoline, jet, and diesel) to run their coking units profitably. With the rising availability of U.S. produced light-sweet crude oil, however, some refineries may choose to cut back on coking, and thus produce less coke. These and other variables lend uncertainty to whether petcoke production will increase, remain steady or decline in the coming years.

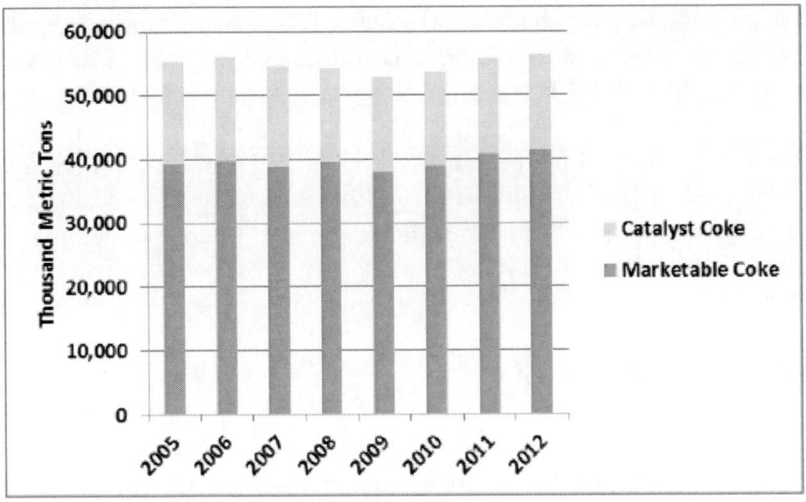

Source: U.S. Energy Information Administration, U.S. Refinery Net Production, http://www.eia.gov/dnav/pet/ pet_pnp_refp2_dc_nus_mbbl_a.htm.

Notes: Catalysts used in various refining processes (e.g., catalytic cracking) become deactivated from the buildup of carbon deposits, In order to reactivate the catalysts; the carbon is burned off and used as a fuel by the refinery. The carbon coke is not recoverable in a concentrated form.

Figure 3. U.S. Refinery Net Petcoke Production; Thousand Metric Tons per Year.

Petcoke Storage Terminals

Refineries temporarily stockpile petcoke on their facilities, but because they generally lack sufficient storage space must transport it regularly to avoid production slowdowns. Typically, coker drums are mounted over railroad tracks so that coke can be discharged directly into open hopper or gondola cars. The rail cars then transport the petcoke to calcining plants or to temporary storage terminals.

A complete accounting of independent terminals that store petcoke exceeds the scope of this report. However, a CRS survey identified at least four companies with petroleum coke as a primary business line: SSM Petroleum Coke LLC, TCP Petroleum coke Corp, DTE Petroleum Coke, LLC, and Kinder Morgan Petroleum Coke Gp LC.[11] SSM Petroleum Coke is an affiliate of Oxbow Carbon LLC (Koch Industries, Inc.). Koch Carbon, LLC specialize in the global sourcing, supply, handling, and transportation of bulk commodities including, but not limited to, petcoke. TCP Petroleum Coke

Corporation is a joint venture between CITGO Petroleum Corporation (CITGO) and RWE Power AG, offering a diversified marketing network to over 30 countries. DTE Petroleum Coke is a subsidiary of DTE Energy, a diversified energy company that includes electric/gas utilities. DTE Energy has reportedly removed the petcoke it stored at its Detroit Bulk Storage site along the Detroit River.[12] Kinder Morgan Petroleum Coke L.P. advertises that it is responsible for handling over 10 million tons of petcoke through several terminals located on the Texas Gulf Coast.

Petcoke Market and Exports

Petcoke competes with both coal and metcoke in the international market. Its comparatively higher heating values makes it an economic substitute for steam coal. However, its granular physical properties may add to the cost of material handling, which is reflected in a discounted price compared with coal in the United States. Petcoke prices track coal prices but at discounts in the range of 15% to 85%.[13] Recently U.S. petcoke price have ranged from 67% to 68% of coal prices.

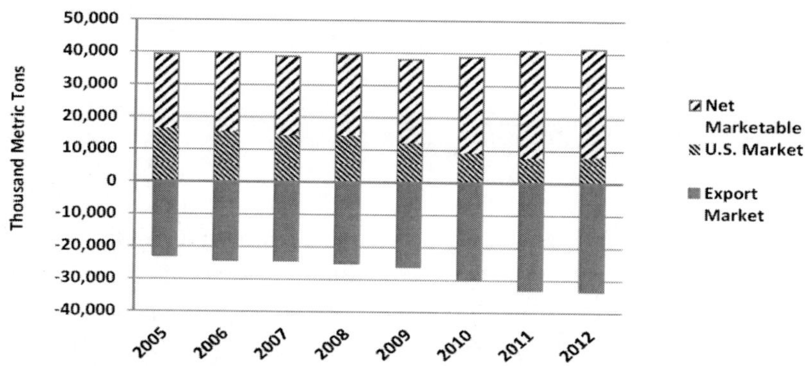

	2005	2006	2007	2008	2009	2010	2011	2012
Net Marketable	39,474	39,818	38,694	39,529	38,105	38,895	40,749	41,376
U.S. Market	16,458	15,577	14,450	14,461	12,185	9,155	7,677	7,952
Export Market	-23,016	-24,241	-24,244	-25,068	-25,920	-29,740	-33,071	-33,424

Source: U.S. Energy Information Administration, U.S. Refinery Net Production, http://www.eia.gov/dnav/pet/ pet_pnp_refp2_dc_nus_mbbl_a.htm, and Petroleum Coke Exports by Destination, http://www.eia.gov/dnav/pet/ pet_move_expc_a_ eppc_eex_mbbl_a.htm.

Figure 4. Net Petcoke Production vs. Exports; Thousand Metric Tons per Year.

U.S. petcoke exports have steadily increased over the last decade, as the U.S. market has given way to increased demand overseas (see **Figure 4**). In 2012, 80% of marketable (i.e., non-catalyst) petcoke was exported. The largest recipients of U.S. petcoke exports in 2012 were China (14%), followed by Japan (11%), Mexico (9%), and Turkey (7%). China's demand has steadily increased during the last decade.

POTENTIAL HEALTH AND ENVIRONMENTAL IMPACTS

The recent increase in coking capacity in the United States has raised concerns over the potential impacts of petcoke on both human health and the environment. Local concerns include air quality hazards, water quality hazards, and potential exposure to toxic substances. These impacts may arise during various stages of petcoke's life-cycle, including its production, handling, storage, transportation, combustion, and use. Broader concerns have been raised about the greenhouse gas (i.e., carbon dioxide) emissions from petcoke combustion. The focus of this report, however, is on the impacts of handling and storage, not on end-use combustion.

EPA's Hazard and Toxicity Characterizations

In recent years, the U.S. Environmental Protection Agency (EPA) has surveyed the potential human health and environmental impacts of petcoke through its High Production Volume (HPV) Challenge Program. The HPV Challenge Program, initiated jointly by EPA, Environmental Defense Fund (EDF), American Petroleum Institute (API), and American Chemistry Council (ACC), was instituted for several purposes, including

1) to collect health and environmental effects data on industrial chemicals produced in the United States in high volumes,
2) to provide the public with basic hazard information that would allow for active participation in environmental decision-making at all levels—local, state, and federal, and
3) to provide EPA with valuable hazard and toxicity information to support its mission of protecting human health and welfare.

Companies have sponsored research into more than 2,200 HPV chemicals, with approximately 1,400 administered directly through the HPV Challenge Program and the remainder administered indirectly through international efforts.[14] API sponsored a testing group for the petcoke category,[15] which produced an analysis in December 2007.[16] This analysis was supplemented by EPA, after stakeholder comments, and published in June 2011.[17]

The findings from EPA's hazard characterization of petcoke are summarized in the following sections.

Environmental Fate

Most chemical analyses of petcoke, as referenced by EPA, find it to be highly stable and non-reactive at ambient environmental conditions.

Due to the extreme conditions under which petcoke is produced, qualities such as melting point, boiling point, vapor pressure, and water solubility exist well outside the range of ambient conditions. If released to the environment, petcoke would not be expected to undergo many of the environmental fate pathways which could lead to environmental risks. Depending on the particle size and density of the material, terrestrial releases of petcoke become incorporated into the soil or transported via wind or surface water flow. If released to the aquatic environment, petcoke incorporates into sediment or floats on the surface, depending on the particle size and density in relation to water. Chemically, petcoke is essentially inert. That is, petcoke does not vaporize into the atmosphere, does not react chemically in the presence of water, and does not react chemically in the presence of light. Furthermore, it is not biodegradable, nor does it bio-accumulate substances—such as toxic chemicals—into its structure.[18]

Environmental Toxicity

Most eco-toxicity analyses of petcoke, as referenced by EPA, find it has a low potential to cause adverse effect on aquatic or terrestrial environments.

The environmental effects of petcoke have been tested along various pathways for exposure in the environment, including both aquatic and terrestrial endpoints in plants and animals. Aquatic and terrestrial toxicity tests have been performed to assess the hazard of petcoke releases to representative aquatic organisms and terrestrial soil-dwelling invertebrates and plants. In these studies, petcoke was found to be non-toxic to terrestrial plants and animals, non-toxic to aquatic animals (both vertebrates and invertebrates), and showed only slight effects on aquatic plants at the exposure levels tested (i.e., studies found slight growth inhibition in freshwater algae).[19] (The exposure levels and durations were conducted in accordance with EPA and Organization for Economic Co-operation and Development (OECD) recommendations, although, presumably, these tests could be re-administered at higher dosages or intervals to assess the effects of greater concentrations.)

Human Health Effects

Most toxicity analyses of petcoke, as referenced by EPA, find it has a low health hazard potential in humans, with no observed carcinogenic, reproductive, or developmental effects. Only animal cases studies of repeated-dose and chronic inhalation have shown respiratory inflammation attributed to the non-specific effects of dust particles rather than the specific effects of petcoke.

Inhalation of and skin contact with petcoke were assessed to be the most likely exposure routes to humans. Most repeated-dose inhalation exposure studies (on rats and primates) found cases of irreversible respiratory effects and significantly increased lung weights. These effects were considered to be non-specific responses of the respiratory tract to high concentrations of dust particles rather than compound specific-induced effects. Petcoke was not found to be carcinogenic via inhalation. No excess skin or visceral cancers were observed in a lifetime skin painting study. Petcoke was not found to produce genetic mutations in bacteria and mammalian cells in standard in vitro toxicity tests or to produce chromosome aberrations of bone marrow in standard in vivo toxicity tests. Petcoke was not found to produce any reproductive or developmental effects following repeated inhalation or exposure to the skin.[20]

Reactivity

Petcoke is generally stable under normal conditions; however, the substance has the potential to become flammable or explosive. Emissions from the combustion—either accidentally or purposefully—of petcoke can have impacts on human health and the environment, including the release of common pollutants, hazardous substances, and greenhouse gases.

When petcoke is combusted, common pollutants and hazardous decomposition products may be produced such as carbon monoxide, carbon dioxide, sulfur dioxide, nitrogen oxides, particulate matter, and heavy metals, depending upon the chemical composition of the feedstock (see **Table 1** for the chemical composition of petcoke). These releases may take place unintentionally, through the natural or unintended combustion of surface or air-borne dust particles, or intentionally, through the combustion of petcoke for electrical power generation or other like purposes.

Petcoke's use as a fuel is criticized because it commonly has higher greenhouse gas emissions relative to the amount of heat it generates when burned. **Table 3** presents potential carbon dioxide (CO_2) emissions for petcoke in comparison to metallurgical coke and several grades of steam coal. When petcoke or coal combust, CO_2 forms from one carbon atom (C) uniting with two oxygen atoms (O).[21] Assuming complete combustion, 1 pound of carbon combines with 2.667 pounds of oxygen to produce 3.667 pounds of carbon dioxide. Petcoke with a carbon content of 90% and a heating value of 14,200 Btu per pound emits about 232 pounds of carbon dioxide per million Btu when completely burned.[22] Comparatively, Powder River Basin coal with a carbon content of 48% and a heating value of 8,800 Btu per pound emits about 202 pounds of carbon dioxide per million Btu when completely burned, or 15% less than petcoke. Because coal has high hydrogen-to-carbon ratio compared to petcoke, part of its energy content comes from the combustion of hydrogen that is emitted as water vapor instead of carbon dioxide.

Table 3. Petcoke vs. Coal: Combustion Emissions

Fuel	Coal Rank	Sulfur %wt.	Btu / lb.	Carbon %wt.	CO2 lbs./ Million Btu
Petcoke	n.a.	1.5–6.0	14,200	89–92	232
Metcoke	Metallurgic	0.4–0.7	12,600	91–92	266
Steam Coal					
Pittsburgh #8	Bituminous	3.0	13,000	73–74	207
Illinois #6	Bituminous	3.9	11,000	60–61	201
Powder River Basin	Sub-bituminous	0.5	8,800	48–49	202

Source: MIT, The Future of Coal Appendices, p. P5.

Notes: Potential carbon dioxide emissions calculated by percent carbon ÷ Btu per pound x 36,670 = pounds (lbs.) of carbon dioxide per million Btu.

Federal Regulatory Requirements

Various aspects of the production, handling, storage, transportation, combustion, and use of petcoke have been addressed at local, state, and federal levels to protect human health and the environment. While some federal statutes address certain environmental impacts of petcoke's life-cycle, most regulatory action and oversight has been undertaken at the state and local levels, generally through facility-specific permitting requirements. With few exceptions, petcoke is not specifically regulated by local, state, or federal codes.[23] Rather, it is petcoke's potential contribution to more general hazards (e.g., air and water quality impacts such as haze, fugitive dust, and stormwater runoff) that is monitored and controlled through various permitting and reporting requirements at the state and local levels. This report focuses on the federal response to petcoke and on the rules that may be implemented during the handling, storage, and transportation phases of petcoke's life-cycle. States may also have their own laws or regulations related to the handling, storage, and transportation of petcoke, specifically, or high-productionvolume industrial substances like petcoke, more generally; a full review of state and local code is beyond the scope of this report.

Regulatory agencies at all levels of government commonly aim to manage the human health and environmental impacts of industrial materials (e.g., petcoke) based upon thorough assessments of their hazardous exposure pathways. Because of its relative inertness, exposure to petcoke is considered to pose few human health and environmental risks. Thus, federal law generally exempts petcoke from classification as either a solid or hazardous waste. Despite these exemptions, petcoke may nevertheless present significant nuisance concerns. A "nuisance" is the unreasonable, unwarranted and/or unlawful use of property, which causes inconvenience or damage to others, either to individuals or to the general public.[24] A nuisance may not violate any regulatory standard or cause demonstrable environmental or health impacts; however, if a nuisance interferes with a person's use of his or her property, it may be the basis for a lawsuit for damages or an injunction. For example, fugitive dust from petcoke storage piles can be deposited on and in nearby waterways, outdoor areas, or residents' homes, leaving a black- or grey-colored residue. This deposition may result in undesirable and unsightly conditions, interfering with residents' comfort and use of their property. Similarly, dust from coke piles can challenge the operations of commercial or industrial facilities, such as pharmaceutical research and production plants, electronics assembly, or fuel cell membrane manufacturing. Dust from nearby coke piles can increase filtration costs or threaten the integrity of strict quality control standards required for such high technology operations.

In light of these issues, the monitoring and management of petcoke at the federal level is summarized in the following sections.

Waste Classifications

Federal law generally exempts petcoke from classification as either a solid or hazardous waste.

The exemption for petcoke from classification as either a solid or hazardous waste stems from the scope of the statutory term "solid waste" as decided in *American Mining Congress v. U.S. EPA.*[25] In that decision, the court held that materials recycled and reused in an ongoing manufacturing or industrial process were not considered to be "discarded," and hence, not considered to be "solid wastes." Furthermore, in 1998, EPA identified a list of petroleum refining wastes that would be subject to federal regulations applicable to the management of hazardous waste established under the Resource Conservation and Recovery Act (RCRA).[26] In this rulemaking, EPA stated that petcoke is not a refining waste, but rather a "co-product" of the refining process.[27] In separate rulemaking, EPA included petcoke among other fuels in its definition of "traditional fuels" (at 40 C.F.R. 241.2). As a result of these determinations, unless or until it is discarded, petcoke would not be subject to federal waste management requirements established under RCRA.

Petcoke would not be subject to the federal cleanup authorities of the Comprehensive Environmental Response, Compensation, and Liability Act (CERCLA, often referred to as Superfund)[28] because of the exclusion of petroleum from the statute. The response authorities of CERCLA specifically apply to the release of hazardous substances, pollutants, or contaminants into the environment. Hazardous substances under CERCLA are broader than hazardous wastes under RCRA and include hundreds of toxic chemicals. However, CERCLA defines a hazardous substance, pollutant, or contaminant to exclude "petroleum, including crude oil or any fraction thereof which is not otherwise specifically listed or designated as a hazardous substance."[29] EPA's interpretation has been that hazardous substances present in petroleum that are naturally occurring or are normally added during the refining process are fractions of the petroleum that would fall within the scope of the exclusion from CERCLA. EPA's position has been that CERCLA may be applied to the cleanup of wastes containing petroleum, if the wastes contain hazardous substances that are not part of the petroleum product itself.[30] Hazardous substances that may leach from a petroleum product into the environment, and therefore no longer be part of the petroleum product, may raise other issues.

Industrial Stormwater Runoff

The handling and storage of petcoke may be regulated under certain provisions of the National Pollutant Discharge Elimination System (NPDES) permit program,[31] as authorized by the Clean Water Act (CWA),[32] if it is determined that runoff from storage sites due to rain or snowmelt has the potential to transport the substance to nearby surface waters.

Activities that take place at industrial facilities, such as material handling and storage, are often exposed to the weather. As runoff from rain or snowmelt comes into contact with these activities, it can pick up pollutants and transport them to a nearby storm sewer system or directly to a river, lake, or coastal water. Recognition of the water quality problems of stormwater runoff led Congress in 1987—when it last comprehensively amended the CWA—to direct EPA to implement a specific permit program for stormwater discharges from industrial sources and municipalities (P.L. 100-4). These stormwater requirements were incorporated into the National Pollutant Discharge Elimination System (NPDES), a comprehensive permit program authorized in Section 402 of the CWA. Under the act, it is illegal to discharge pollutants from point sources (e.g., industrial plant pipes, sewage treatment plants, or storm sewers) into the nation's waters without a permit. NPDES permits are the fundamental compliance and enforcement mechanism of the law. EPA manages the NPDES stormwater program in four states (Idaho, Massachusetts, New Hampshire, and New Mexico), plus the District of Columbia and most U.S. territories, and has delegated that authority to the remaining 46 states and the Virgin Islands. Therefore, the vast majority of industrial and other facilities obtain NPDES permit coverage for stormwater discharge through their state. Petroleum refining facilities are one of several categories of facilities that are specifically covered under the CWA stormwater regulatory program.

Common requirements for coverage under an industrial stormwater permit include development of a written stormwater pollution prevention plan (SWPPP), implementation of control measures, and submittal of a request for permit coverage, usually referred to as the Notice of Intent (NOI). The SWPPP is a written assessment of potential sources of pollutants in stormwater runoff and control measures that will be implemented at the facility to minimize the discharge of these pollutants in runoff from the site. These control measures include site-specific best management practices, maintenance plans, inspections, employee training, and reporting. The procedures detailed in the SWPPP must be implemented by the facility and updated as necessary, with a copy of the SWPPP kept on-site. The industrial stormwater permit also requires collection of visual, analytical, and compliance monitoring data to determine the effectiveness of implemented best management practices. Stormwater permits are valid for up to five years and must be renewed.

Best management practices for the prevention of industrial stormwater runoff include ensuring adequate storage facilities and equipment, spill detection and repair, and employee training. Many environmental agencies, including EPA, provide extensive summaries of best management practices.[33]

Fugitive Dust

The handling, storage, and transportation of petcoke may create instances of reduced air quality due to weather or activity related releases of fugitive dust into the atmosphere. Most of the impacts of fugitive dust are localized; and thus, much of the regulatory oversight is implemented at the local and state level and generally takes the form of a fugitive dust control program.

Facilities may be required by state or local agencies to develop a fugitive dust control program for many reasons. State and local agencies, based on their own air emission measurements, their own code of regulations, environmental consent orders, or complaints of nuisance, may require a fugitive dust program from any facility if it processes, uses, stores, transports, or conveys bulk materials from a highly emitting dust source. Further, these programs are often a necessary component to any air permitting requirements at the state and local level, including permits to install, operate, or decommission a facility. At the federal level, Clean Air Act (CAA)[34] National Ambient Air Quality Standards (NAAQS) have been set nationwide by EPA for particulate matter (PM). [35] NAAQS are standards for outdoor (ambient) air that are intended to protect public health and welfare from harmful concentrations of pollution. If fugitive dust generation is determined to be an issue at a facility that produces, handles, stores, transports, or uses petcoke, and if the facility is situated in an area that is identified by the EPA as "nonattainment" for PM NAAQS, then it may be possible for state authorities to ask the facility to report on and manage its fugitive dust emissions—if it is not doing so already—within the context of their State Implementation Plans (SIPs).

Whether petcoke storage is considered a significant source of PM depends a number of factors, including how the coke is handled (e.g., number of drops), individual petcoke particle sizes and the size of the overall petcoke piles, as well as the storage method. In some cases, petcoke may have been processed through pulverization that generates dust which could be monitored and controlled at PM2.5 (less than 2.5 microns). PM2.5 can produce greater health impacts because it is more respirable than "coarse" PM which is larger than 2.5 microns. PM that is greater than about 10 microns is generally considered less of a health risk because it is less respirable. As illustrated in **Figure 1**, most forms of petcoke are comprised of granules orders of magnitude larger than PM2.5, and are not likely respirable, but may pose a nuisance concern. Also, in some cases, petcoke storage may be ephemeral because markets support frequent elimination of stored inventories.

The management of fugitive dust commonly involves the submission of a fugitive dust plan to state or local agencies. These plans would include an analysis of the quantity and opacity of fugitive dust from the facility; a determination of the type of fugitive dust control methods that would be the most effective, taking into account the quantity, moisture content, specific gravity, and particle size distribution of the bulk materials on-site; an assessment of the type of control technologies, methods, and equipment to be implemented or installed, and the schedule for implementation or installation; and a report on the level of recordkeeping and maintenance requirements for activities that are implemented under the dust program. Fugitive dust plans commonly set out an operating program designed to significantly reduce emissions to the lowest level that a particular source is capable of achieving by the application of control technology that is both reasonably available and based on technological and economic feasibility. The requirement for fugitive dust plans for a given facility and the plan's enforcement remain at the discretion of the state and local agencies.

Best management practices for the prevention of fugitive dust include ensuring adequate storage facilities and equipment, emission detection and repair, and employee training. Many environmental agencies, including the U.S. EPA, provide extensive summaries of best management practices.[36]

Regulatory Standards for Petcoke Combustion in Power Generation

The combustion of petcoke, and the resulting emissions from this combustion, may be regulated under several different provisions in the CAA and the CWA. When petcoke is used for industrial or power generating purposes, emissions from its combustion are regulated under the standards set on the respective facilities. For example, some of the federal regulations which may be implemented by the combustion of petcoke at industrial or power generating facilities include EPA's New Source Performance Standards for Electricity Generating Units, Clean Air Interstate Rule (CAIR) for NO_x and SO_2, CAA Title V Permitting Requirements, GHG Reporting Program, Steam Electric Effluent Guidelines, recently finalized Mercury and Air Toxics Standards/Utility MACT, and proposed Coal Combustion Waste Rules, as well as the various state rules under State Implementation Plans.

A full description of these requirements is beyond the scope of this report.

CONCLUSION

Petcoke production depends on the crude oil demand of operating refineries that in tum depend on the availability of heavy and light crude oils and their comparative cost advantages. The domestic production of petcoke may increase as U.S. refineries continue to add coking capacity to take advantage of competitively priced petroleum produced from Canada's oil sands and other heavy crude oil sources. Conversely, as U.S. light crude oil production increases, U.S. coking refineries may find an economic advantage in switching to lighter crude oils and idle their coking units. If that occurs, the production and export of U.S. petcoke may reverse. Community stakeholder and regulator concerns about fugitive dust emission into the air and stormwater runoff into waterways are likely to continue in situations where there is not sufficient mitigation and abatement. In some states, permit conditions have been imposed to mitigate the emissions from petcoke storage and handling. The specific permit conditions (e.g., enclosed conveyors and storage silos) are generally based on best management practices as determined by state regulators. At the federal level, Clean Air Act (CAA) National Ambient Air Quality Standards (NAAQS) for outdoor (ambient) air continue to protect public health and welfare from harmful concentrations of particulate matter pollution. If states determine that fugitive dust generation is an issue at a facility that produces, handles, stores, transports, or uses petcoke, and if the facility is situated in an area that is identified by the EPA as "nonattainment" for PM NAAQS, then state authorities may ask the facility to report on and manage its fugitive dust emissions—if it is not doing so already—within the context of their State Implementation Plans (SIPs). States and localities may also have their own regulatory standards for fugitive dust, independent of whether the area is in nonattainment of federal PM NAAQS.

In light of these concerns, industry, regulators, and compliance officers have shown a continued interest in impact assessment and best practices related to the storing, containing, and managing of petcoke. Two bills have been introduced in the 113th Congress regarding petcoke: H.R. 2298, the Petroleum Coke Transparency and Public Health Study Act (introduced 6/6/2013), and S. 1388, Petroleum Coke Transparency and Public Health Study Act (introduced 7/30/2013). Each would require the Secretary of Health and Human Services, in consultation with the Administrator of the Environmental Protection Agency, to conduct a study on the public health and environmental impacts of the production, transportation, storage, and use of petcoke.

APPENDIX. PETROLEUM REFINING AND PETCOKE PRODUCTION

Petroleum refineries use several key processes to convert crude oil systematically into refined products; these include atmospheric distillation, hydrocracking, hydrotreating, reforming, and ultimately coking. The refinery's atmospheric distillation column initially separates crude oil into lighter streams of hydrocarbons based on their boiling temperatures. The gasoline-range of petroleum distillates condense at the top of the column. Middle distillate fuels (kerosene, jet, and diesel fuels) condense in the middle of the column. The heavier-still range of gas oils condense lower in the column. Residuum, a heavy tar-like material figuratively referred to as the "bottom of the barrel," has such a high boiling temperature that it remains at the bottom of the column.

In order to produce more gasoline, refineries "crack" the heavier distillation products into the gasoline range with heat, pressure, hydrogen, and catalysts. Hydrotreating removes elemental sulfur from gasoline and middle-distillate fuels through a reaction with hydrogen gas.

Coking dates back to the late 1920s, but became an important process for U.S. refineries during the 1980s and 1990s. During this time, refineries faced a dwindling supply of light sweet crude oils favored for making gasoline and distillate fuels. They began switching to increasingly more available, heavy-sour crude oils. The resid that remained after refining heavier crudes initially found use as "ship's bunker fuel" and as boiler fuel in electric power plants. With the implementation of Clean Air Act regulations, power plants switched from boiler fuel to cleaner burning natural gas. During the same era, the demand for gasoline increased, and refineries began adding coking to convert the "resid" into motor fuels.

Coking initially converts petroleum residuum into lighter range hydrocarbons; low-Btu gas that can serve as a fuel in refinery operations; and "green coke."

Refineries commonly employ one of three types of coking processes:

- *delayed coking*—a thermal cracking process that converts residuum into gasified products streams and concentrated carbon coke. It is called "delayed coking" because cracking takes place in a coke drum rather than in a furnace or reactor. The residuum is heated in a furnace first, and then fed into the bottom of the coke drum. The "cracked" light products are drawn off at the top of the drum and sent to a fractionator which separates out gasoline, naphtha, gas oil, and lighter products. The drums are "de-coked" by hydraulic or mechanical cutting processes. In delayed coking, one coking drum is filled while a second is decoked (emptied). First commercialized in 1928, delayed coking predominates among U.S. refineries that process heavy crude oil. See **Figure A-1**.

- *flexi-coking*—a continuous fluidized-bed thermal cracking process integrated with coke gasification. It converts most of the carbon coke to carbon monoxide (CO), which is then mixed with carbon (C2) and lighter hydrocarbons to produce a low quality fuel gas. The process was commercialized in 1976. See **Figure A-2**.
- *fluid coking*—a variation on flexi-coking that uses a cyclone to separate the coke. The process was commercialized in 1954.

Source: Shell.

Figure A-1. Shell Delayed Coking Unit; Puget Sound Refinery.

Source: ExxonMobil Research & Engineering Company, Upgrading of Heavy Oils with Flexicoking.

Figure A-2. ExxonMobil Flexi-Coking Unit; Rotterdam, The Netherlands.

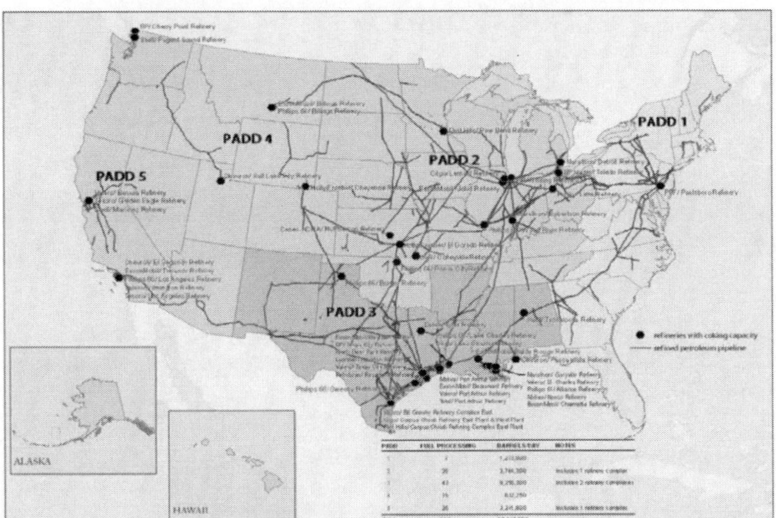

Source: Prepared for CRS by the Library of Congress.
Notes: See Table A-1 for a list of refineries.

Figure A-3. U.S. Refineries with Coking Capacity, by Petroleum Adminstration for Defense Districts (PADD).

Independent processors convert the green coke into to either fuel grade or anode grade coke depending upon the crude oil refined and the coking process used.

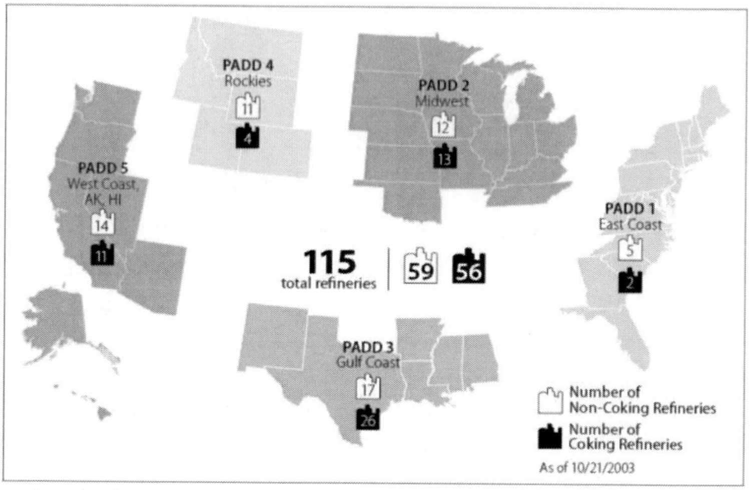

Source: Compiled by CRS from various sources.

Figure A-4. Coking Refineries by PADD.

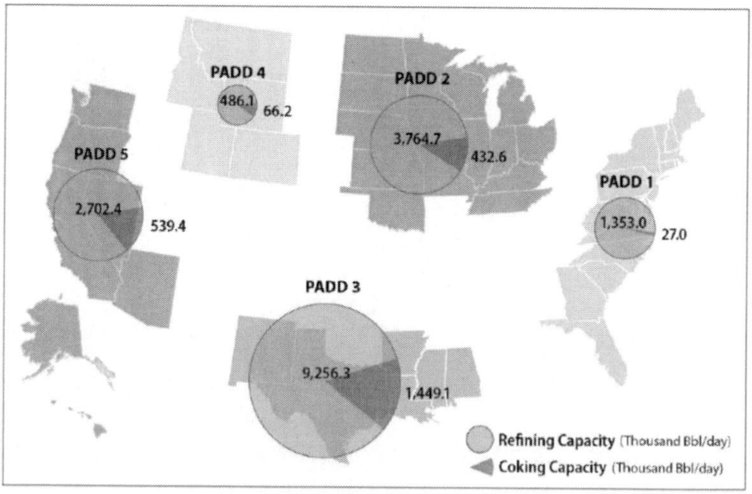

Source: Compiled by CRS from various sources.

Notes: Coking capacity denotes the throughput capacity to process petroleum resid.

Figure A-5. Refining and Coking Capacity by PADD.

Table A-1. U.S. Refineries with Coking Capacity Advertised Refinery Capacity in Barrels per Day (Bbl/D)

PADD	St	City	Zip	Facility	Bbl/D	Congressional District
1	NJ	Paulsboro	08066	PBF/ Paulsboro Refinery	180,000	1st New Jersey
1	DE	Delaware	19706	PBF/Delaware City	190,000	At large Delaware
2	IL	Roxana	62084	Phillips 66/ Wood River Refinery	306,000	12th Illinois
2	IL	Drummond	60410	ExxonMobil/ Joliet Refinery	250,000	16th Illinois
2	IL	Robinson	62454	Marathon/ Robinson Refinery	206,000	15th Illinois
2	IL	Lemont	60439	Citgo/ Lemont Refinery	167,000	3rd Illinois
2	IN	Whiting	46394	BP/ Whiting Refinery	413,000	1st Indiana
2	KS	El Dorado	67042	Holly-Frontier/ El Dorado Refinery	135,000	4th Kansas
2	KS	Coffeyville	67337	CVR Coffeyville Refinery	115,000	2nd Kansas
2	KS	McPherson	67460	Cenex-NCRA/ McPherson Refinery	85,000	1st Kansas
2	MI	Detroit	48217	Marathon/ Detroit Refinery	106,000	13th Michigan
2	MN	Rosemont	55068	Flint Hills/ Pine Bend Refinery	320,000	2nd Minnesota
2	OH	Oregon	43616	BP-Husky/ Toledo Refinery	160,000	9th Ohio
2	OH	Lima	45804	Husky/ Lima Refinery	155,000	4th Ohio
2	OK	Ponca City	74601	Phillips 66/ Ponca City Refinery	187,000	3rd Oklahoma
3	AL	Tuscaloosa	35401	Hunt/ Tuscaloosa Refinery	72,000	7th Alabama
3	LA	Baton Rouge	70805	ExxonMobil/ Baton Rouge Refinery	503,500	2nd Louisiana
3	LA	Garyville	70051	Marathon/ Garyville Refinery	490,000	2nd Louisiana
3	LA	Lake Charles	70601	Citgo/ Lake Charles Refinery	425,000	3rd Louisiana
3	LA	Norco	70079	Valero/ St. Charles Refinery	270,000	6th Louisiana
3	LA	Belle Chasse	70037	Phillips 66/ Alliance Refinery	247,000	1st Louisiana
3	LA	Westlake	70669	Phillips 66/ Lake Charles Refinery	239,000	3rd Louisiana

Table A-1. (Continued)

PADD	St	City	Zip	Facility	Bbl/D	Congressional District
3	LA	St. Charles Parrish	70079	Motiva/ Norco Refinery	234,700	6th Louisiana
3	LA	Chalmette	70043	ExxonMobil/ Chalmette Refinery	192,500	1st Louisiana
3	MS	Pascagoula	39581	Chevron/ Pascagoula Refinery	330,000	4th Mississippi
3	TX	Port Arthur	77641	Motiva/ Port Arthur Refinery	600,000	14th Texas
3	TX	Baytown	77520	ExxonMobil/ Baytown Refinery	573,000	36th Texas
3	TX	Texas City	77590	Marathon/ Texas City Refinery	475,000	14th Texas
3	TX	Beaumont	77703	ExxonMobil/ Beaumont Refinery	365,000	14th Texas
3	TX	Deer Park	77536	Shell/ Deer Park Refinery	340,000	36th Texas
3	TX	Corpus Christi	78407	Valero/ Bill Greehy Refinery Complex East	325,000	27th Texas
3	TX	Port Arthur	77641	Valero/ Port Arthur Refinery	310,000	14th Texas
3	TX	Houston	77017	Lyondell/ Houston Refinery	268,000	29th Texas
3	TX	Sweeny	77463	Phillips 66/ Sweeny Refinery Complex	247,000	14th Texas
3	TX	Texas City	77590	Valero/ Texas City Refinery	245,000	14th Texas
3	TX	Port Arthur	77642	Total/ Port Arthur Refinery	174,000	14th Texas
3	TX	Corpus Christi	78047	Citgo/ Corpus Christi Refinery East & West Plant	165,000	27th Texas
3	TX	Corpus Christi	78408	Flint Hills/ Corpus Christi Refining Complex East Plant	150,000	27th Texas
3	TX	Borger	79007	Phillips 66/ Borger Refinery	146,000	13th Texas
3	TX	Pasadena	77506	Petrobras/ Pasadena Refinery	100,000	29th Texas
3	TX	Tyler	75702	Delek/ Tyler Refinery	60,000	1st Texas
4	MT	Billings	59101	ExxonMobil/ Billings Refinery	60,000	At Large Montana
4	MT	Billings	59101	Phillips 66/ Billings Refinery	58,000	At Large Montana

PADD	St	City	Zip	Facility	Bbl/D	Congressional District
4	UT	Salt Lake City	84116	Chevron/ Salt Lake City Refinery	45,000	2nd Utah
4	WY	Cheyenne	82007	Holly-Frontier/ Cheyenne Refinery	52,000	At Large Montana
5	CA	Paramount	90723	Alon/ California Refineries	70,000	40th California
5	CA	El Segundo	90245	Chevron/ El Segundo Refinery	290,000	33rd California
5	CA	Benicia	94510	Valero/ Benicia Refinery	170,000	5th California
5	CA	Martinez	94553	Tesoro/ Golden Eagle Refinery	166,000	5th California
5	CA	Martinez	94553	Shell/ Martinez Refinery	165,000	5th California
5	CA	Torrance	90509	ExxonMobil/ Torrance Refinery	150,000	43rd California
5	CA	Carson	90745	Phillips 66/ Los Angeles Refinery	139,000	44th California
5	CA	Wilmington	90744	Valero/ Wilmington Refinery	135,000	44th California
5	CA	Wilmington	90744	Tesoro Los/ Angeles Refinery	97,000	44th California
5	WA	Blaine	98230	BP/ Cherry Point Refinery	230,000	1st Washington
5	WA	Anacortes	98221	Shell/ Puget Sound Refinery	145,000	2nd Washington

Source: Various.

Notes: Alon operates three units in Bakersfield, Paramount, and Long Beach, CA, as one refinery, but the delayed coker is reported as inactive.

End Notes

[1] Pet Coke Consulting LLC, http://www.petcokeconsulting.com/primer/index.html.

[2] Thermal coal is sometimes called "steam coal" because it is used to fire boiler plants that produce steam for electricity generation and industrial uses.

[3] Anthony Pavone, "Converting Petroleum Coke to Electricity," Proceedings from the 14th National Industrial Energy Technology Conference, Houston, TX, April 22-23, 1992, http://repository.tamu.edu/bitstream/handle/1969.1/92212/ ESL-IE-92-04-47.pdf.

[4] Bituminous coals are mined throughout the eastern United States range but generally have higher sulfur and nitrogen contents than western coals.

[5] Oxbow, *Metallurgical Coke*, http://www.oxbow.com/Products_Industrial_Materials_ Metallurgical_Coke.html. Also see Grande Cache Coal, *Met Coke 101*, http://www. gccoal.com/about-us/met-coal

[6] For further background on the refining industry, see CRS Report R41478, *The U.S. Oil Refining Industry: Background in Changing Markets and Fuel Policies*, by Anthony Andrews et al.

[7] The U.S. Energy Information Administration reported that U.S. refinery net production of petroleum coke in 2012 was 310,481 thousand barrels (U.S. Refinery Net Production, http://www.eia.gov/dnav/pet/ pet_pnp_refp2_dc_nus_mbbl_a.htm). 1 metric ton is the equivalent of 5.51 barrels.

[8] American Iron and Steel Institute, *Annual Statistical Report*, 2012, p.80. Production and Consumption of Coke.

[9] U.S. Energy Information Administration, *Monthly Energy Review*, Coal, June 2013, http://www.eia.gov/totalenergy/ data/monthly/index.cfm#coal.

[10] U.S. Energy Information Administration, Refinery Utilization and Capacity, http://www.eia.gov/dnav/pet/ pet_pnp_unc_dcu_nus_a.htm.

[11] A search through http://www.Manta.com produced six U.S. companies matching "petroleum coke."

[12] "Pet Coke Piles Along Detroit River Clear Away," *CBS Detroit*, August 27, 2013, http://detroit.cbslocal.com/2013/ 08/27/pet-coke-piles-along-detriot-river-cleared-away/.

[13] Personal communication with Mike Stewart, Jacobs Consultancy/Petroleum Coke Quarterly.

[14] U.S. Environmental Protection Agency's High Production Volume (HPV) Challenge Program, http://www.epa.gov/ chemrtk/index.htm.

[15] U.S. Environmental Protection Agency, *Robust Summaries & Test Plans: Petroleum Coke*, http://www.epa.gov/ chemrtk/pubs/summaries/ptrlcoke/c12563tc.htm.

[16] The American Petroleum Institute Petroleum HPV Testing Group, Petroleum Coke Category Analysis and Hazard Characterization, submitted to EPA December 28, 2007, http://www.epa.gov/hpv/pubs/summaries/ptrlcoke/ c12563rr2.pdf.

[17] U.S. Environmental Protection Agency, Screening-Level Hazard Characterization, Petroleum Coke Category, June 2011, http://www.epa.gov/chemrtk/hpvis/hazchar/Category_ Petroleum%20Coke_June_2011.pdf.

[18] Petcoke's volatilization is negligible, its rate of hydrolysis is negligible, and its rate of atmospheric photo-oxidation is negligible. Neither biodegradation nor bioaccumulation is applicable.

[19] Environmental toxicity studies referenced by the EPA analysis include Wildlife International, Ltd., *Petroleum Coke: A 96-Hour Static-Renewal Acute Toxicity Test with the Fathead Minnow (Pimephales promelas), Final Report,* Project No. 472A-1134, 2006; Wildlife International, Ltd., *Petroleum Coke: A 48-Hour Static-Renewal Acute Immobilisation Test with the Cladocern (Daphnia magna), Final Report*, Project No. 472A-112, 2006; Wildlife International, Ltd., *Petroleum Coke: A 96-Hour Toxicity Test with the Freshwater Alga (Selenastrum capricornutum), Final Report,* Project No. 472A-114, 2006; Wildlife International, Ltd., *Petroleum Coke: A 21-Day Toxicity Test to Determine the Effects of the Test Substance on Seedling Emergence and Growth of Terrestrial Plants, Final Report,* Project No. 472- 102, 2006; Wildlife International, Ltd., *Petroleum Coke: A 14-Day Acute Toxicity Test with the Earthworm (Eisenia fetida), Final Report,* Project No. 472-101, 2006.

[20] Toxicity studies referenced by the EPA analysis include Huntingdon Life Sciences, Study No. 03-4246, "Petroleum Coke: Reproduction/Developmental Toxicity Screening Study in Rats via Nose-Only Inhalation Exposures," 2006; Klonne, D. R., Burns, J. M., Halder, C. A., Holdsworth, C. E. and Ulrich, C. E., "Two Year Inhalation Study of Petroleum Coke in Rats and Monkeys." *American Journal of Industrial Medicine*, 11:375-389, 1987; and IRDC (International Research & Development Corporation), "Chronic Inhalation Toxicity Study of Petroleum Coke (Delayed Process) in Rats and Monkeys," API Publication number 32-30234, 1985.

[21] B. D. Hong and E. R. Slatick, *Carbon Dioxide Emission Factors for Coal*, U.S. Energy Information Administration, http://www.eia.gov/coal

[22] Potential carbon dioxide emissions can be calculated by use of the following formula: percent carbon ÷ Btu per pound x 36,670 = pounds (lbs.) of carbon dioxide per million Btu.

[23] In a survey of state statutes, conducted by the U.S. Library of Congress's Law Library for this report on August 9, 2013, California was found to be the only state which has passed laws to directly manage the environmental impacts from the handling, storage, and transportation of petcoke. California State Code on petcoke included California Code– HSC Section 40459 (requirements for enclosing piles of petcoke when storing the substance prior to shipment), http://www.leginfo.ca.gov/cgi-bin/displaycode?section=hsc&group=40001-41000&file=40440-40459, and California Code–VEH Section 23114 (requirements for the transportation of petcoke by vehicle), http://www.leginfo.ca.gov/cgibin/displaycode?section=veh&group=23001-24000&file=23100-23135.

[24] A nuisance may be either a private nuisance or a public nuisance. An activity constitutes a *private* nuisance if it is a substantial and unreasonable invasion of another's interest in the private use and enjoyment of land, without involving trespass. Private nuisance actions are brought by the aggrieved landowner. An activity is a *public* nuisance if it creates an "unreasonable" interference with a right common to the general public. Unreasonableness may rest on the activity significantly interfering with, among other things, public health and safety. Public nuisance cases are usually brought by the government rather than private entities, but may be brought by the latter if they suffer special injury.

[25] *American Mining Congress v. U.S. EPA*, 824 F.2nd 1177 (D.C. Cir. 1987). The court held that the EPA exceeded its authority by amending its definition of "solid waste" under the Resource Conservation and Recovery Act (RCRA) to include secondary materials destined for reuse within an industry's ongoing production process. The court held that EPA's interpretation is contrary to RCRA's plain language (§ 1004(5) defines solid waste to include "discarded material"), and that EPA's inclusion of materials retained for immediate use as discarded material strains the everyday usage of that term.

[26] 42 U.S.C. 6901 et seq. For further discussion of the authorities of RCRA, see CRS Report RL30798, *Environmental Laws: Summaries of Major Statutes Administered by the Environmental Protection Agency*, coordinated by David M. Bearden.

[27] See U.S. Environmental Protection Agency, Final Rule, "Hazardous Waste Management System; Identification and Listing of Hazardous Waste; Petroleum Refining Process Wastes; Land Disposal Restrictions for Newly Identified Wastes; and CERCLA Hazardous Substance Designation and Reportable Quantities," August 6, 1998, 63 *Federal Register* 42110. "The coke product itself may best be characterized as a co-product of the coking operation, while the principal products are the light ends that are returned to the refining process. Thus, the Agency is affirming that the conventional coking operation is a production process, resids are normal feedstocks to this process and petroleum coke is a legitimate fuel product." Id, at page 42121.

[28] 42 U.S.C. 9601 et seq. For further discussion of the authorities of CERCLA, see CRS Report R41039, *Comprehensive Environmental Response, Compensation, and Liability Act: A Summary of Superfund Cleanup Authorities and Related Provisions of the Act*, by David M. Bearden.

[29] The definition of the term "hazardous substance" in Section 101 (14) of CERCLA is codified at 42 U.S.C. 9601(14). The definition of the terms "pollutant" and "contaminant" in Section 101(33) of CERCLA is codified at 42 U.S.C. 9601(33).

[30] U.S. Environmental Protection Agency, Office of General Counsel, *Scope of the CERCLA Petroleum Exclusion Under Sections 101(14) and 104(a)(2)*, July 31, 1987. Section 104(a)(2) of CERCLA, as originally enacted in 1980, defined the term "pollutant or contaminant." Section 101(f) of the Superfund Amendments and Reauthorization Act of 1986 re-designated the definition of this term in Section 101(33) of CERCLA, cited above. The full text of the 1987 guidance is available at http://www.epa.gov/compliance

[31] For further discussion on the NPDES Permit Program, see CRS Report 97-290, *Stormwater Permits: Status of EPA's Regulatory Program*, by Claudia Copeland, as well as U.S. Environmental Protection Agency, "National Pollutant Discharge Elimination System," http://cfpub.epa.gov/npdes/.

[32] 33 U.S.C. §§1251-1387. For further discussion of the authorities of CWA, see CRS Report RL30030, *Clean Water Act: A Summary of the Law*, by Claudia Copeland.

[33] An extensive list of best management practices for stormwater runoff, included under the general categories of "good housekeeping practices," "minimize exposure," "erosion and sediment control," and "management of runoff," can be found in U.S. Environmental Protection Agency, *Industrial Stormwater, Fact Sheet Series, Sector C: Chemical and Allied Products Manufacturing and Refining,* U.S. EPA Office of Water, EPA-833-F-06-018, December 2006, http://www.epa.gov/npdes/pubs/sector_c_chemical

[34] 33 U.S.C. §§1251-1387. For further discussion of the authorities of CWA, see CRS Report RL30030, *Clean Water Act: A Summary of the Law*, by Claudia Copeland.

[35] For further discussion of particulate matter, see CRS Report RL34762, *The National Ambient Air Quality Standards (NAAQS) for Particulate Matter (PM): EPA's 2006 Revisions and Associated Issues*, by Robert Esworthy.

[36] An extensive list of emission control techniques for fugitive dust can be found in U.S. Environmental Protection Agency, *Fugitive Dust Background Document and Technical Information Document for Best Available Control Measures,* U.S. EPA Office of Air and Radiation, EPA-450/2-92-004, September 1992, http://www.epa.gov/oaqps001/lead/pdfs/1992_09_fugitive_dust.pdf.

In: U.S. Petroleum Refining
Editor: Roger M. Waters

ISBN: 978-1-63463-336-9

Chapter 2

SMALL REFINERIES AND OIL FIELD PROCESSORS: OPPORTUNITIES AND CHALLENGES*

Anthony Andrews, Richard K. Lattanzio, Robert Pirog, James D. Werner and Brent D. Yacobucci

SUMMARY

The last refinery constructed in the United States opened in 1977. Since the mid-1980s, some 150 have closed as part of an industry-wide consolidation. Over the same time, the remaining refineries expanded their operational capacity by 23% to keep up with increasing demand. Current U.S. refining capacity appears to satisfy if not exceed demand as the increasing export of refined petroleum products would seem to suggest. Notwithstanding the current surplus capacity, opportunities for new refineries appear to have emerged as the result of the rise in production of U.S. light-sweet crude oil from unconventional resources such as North Dakota's Bakken and Texas' Eagle Ford formations. These new resources have revitalized some refinery operations that formerly depended on imported light crude oil, thereby making the smaller refineries more competitive with large refineries that process more widely available heavy-sour crude oil. Rising domestic crude oil production has

* This is an edited, reformatted and augmented version of a Congressional Research Service publication R43682, prepared for Members and Committees of Congress, dated August 11, 2014.

not only led U.S. refineries to export their petroleum products, but has led some oil producers to attempt to bypass refining and export crude oil condensates directly. However, Department of Commerce regulations currently restrict crude oil exports. Whether condensates fall under a refined product classification or crude oil remains an unresolved issue, and one that Congress may choose to take up. Congress may also consider whether small businesses face inherent disadvantages in entering an industry dominated by large complex refinery operators.

While some arguments remain for new refinery construction, some argue that economic and regulatory barriers hinder new construction. Small refining enterprises may suffer certain economic disadvantages in an industry driven by the economy of scale. Furthermore, opportunities may depend on geographic location, regional markets, and source of crude oil. The barriers from regulatory requirements remain less clear. Export restrictions certainly affect potential product markets that could support new refining capacity. Whether these markets would spur new refinery construction is a separate question.

The history of refinery expansions suggests that environmental requirements do not generally pose a significant barrier to refinery construction, but local challenges vary (e.g., locations with currently unhealthy air quality). Small refineries face many of the same economic, market, and environmental factors that affect large refineries, but they may also benefit from exemptions in complying with certain federal regulations. As with any industrial facility involving potential air emissions, water use, and waste generation, certain environmental permitting and regulatory requirements apply to the construction and operation of refineries. The specific requirements, and the degree of difficulty or ease in meeting them, will depend not only on the nature of the project, but perhaps more importantly on the location of the project.

Small business opportunities in refining may strongly depend on a number of factors, including regional demand for refined products, available crude oil supplies, and unique regional environmental regulations, among others. A "small business" in the oil refining business is defined differently in different statutes. In addition, some fuels are easier to produce than others are. While evaluating opportunities is beyond the scope of this report, some key opportunities have recently emerged. Perhaps the most discussed potential opportunity for constructing a small topping or hydroskimming refinery is in the mid-continent region (North Dakota, Colorado) to process these new sources of light sweet crude. Other opportunities may lie in installing condensate splitter and stabilizer plants to handle North Dakota and Texas expanding production of light sweet crude oil and condensate.

INTRODUCTION

The last new refinery to open in the United States finished construction in 1976. That record is about to fall with the construction of a small refinery scheduled to open late this year in North Dakota. Since the mid-1980s, though, some 150 or more refineries closed. However, the refineries that remained open expanded their operational capacity to keep up with the increasing demand for refined petroleum products. Current U.S. refining capacity (roughly 18 million barrels per day) appears to satisfy if not exceed demand as the increasing export of refined petroleum products would seem to suggest. The refining industry even shows signs of expanding to take advantage of less expensive U.S. crude oil and export more products. The perception remains, nevertheless, that the United States needs new refineries but economic and regulatory barriers stand in the way of their construction. Thus far, the industry has managed to negotiate regulatory hurdles in expanding refinery capacity, an option preferred over new construction and driven by the economy of scale.

In general, new entrants into refining may find steep competition with existing operators. However, for some "teapot" refineries, the door is wide open, and the cost of admission is comparatively little when processing light-sweet crude oil or condensate, whether in the mid-continent or elsewhere. Some may argue that there may be only a short window of opportunity to get in and then get out. A change in crude oil prices could quickly disadvantage them.

Small refineries do face many of the same economic, market, and environmental factors that affect large refineries, but they may also benefit from exemptions in complying with certain federal regulations. Further, some federal policies have had the effect of displacing petroleum products in the market place, and introduced refineries to direct challenges, such as the Renewable Fuel Standard.

We begin this report with a brief discussion on the state of refining in the Unites States, paying particular attention to regulatory criteria that define refining (as this has taken on new significance), the threshold between small and large refineries, and potential opportunities for small businesses. In the discussion on "Economic Challenges", we examine the pace of refinery construction and capacity expansion, and the profitability of operating refineries. Current economic challenges for the industry also include competition from ethanol and biodiesel producers, many of which are considered small businesses; the potential displacement of refined petroleum products by these alternative fuels, as mandated by the Renewable Fuel

Standard; and the erosion in demand for refined petroleum products, due to both consumer and government-mandated trends toward increased automotive fuel efficiency.

Like any other industrial facility, refineries are subject to certain environmental laws that regulate their construction and operation. In the discussion on Environmental Permitting and Regulatory Requirements, we summarize some of the major environmental requirements associated with refinery construction and operations, few of which are unique to refineries compared to any industrial operation with similar potential emissions. This discussion focuses on air emissions control requirements, often regarded as the most significant challenge in permitting the expansion of existing refineries or constructions of new refineries.

BACKGROUND

By CRS's recent count, some 115 refineries/refinery complexes process roughly 18 million barrels per day (Mb/d) of crude oil in the United States.[1] All have atmospheric distillation in common, which involves heating crude oil in a furnace then condensing it in an atmospheric distillation tower (or crude unit)—the tall, narrow columns that give a refinery its distinctive skyline (see *Figure 1*).

Refineries vary in complexity, and can include, for the purposes of this report, stabilizers, condensate splitters, simple topping plants and hydroskimmers to more complex cracking and ultimately coking plants. Coking has become an increasingly important capability of the U.S. refining industry with nearly three-quarters of the fuel-producing refineries having some capacity to convert heavy crude oils into refined products.

Beginning in the late 1970s, U.S refineries began facing a dwindling supply of light-sweet crude oils favored for making motor fuels. Many refineries began switching to increasingly available heavy-sour crude oils, and began adding "cracking" and "coking" processes to convert petroleum "resid" into high value motor fuels. The resid that remained after atmospheric distillation found earlier use as low value "ship's bunker fuel" and as boiler fuel in electric power plants. Cracking and coking processing upgrades came at considerable investment costs. Coking dates back to the late 1920s, but became the defining characteristic of the U.S. refinery during the 1980s and 1990s.

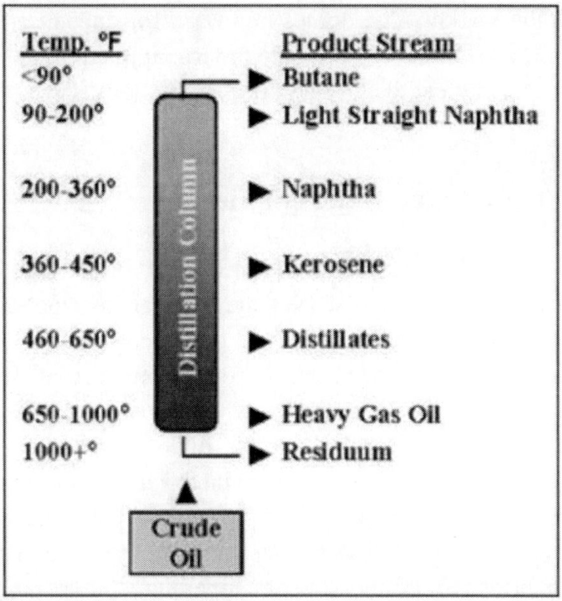

Source: Prepared by CRS.
Notes:

Refinery Type	Processes
Topping	Atmospheric distillation to separate crude oil into constituent petroleum product, produces naphtha, and middle distillate fuels, and resid, but no gasoline
Hydroskimming	Atmospheric distillation, naphtha reforming and desulfurization process to run light sweet crude and produce gasoline
Cracking	Adds vacuum distillation and catalytic cracking process to run light sour crude to produce light and middle distillate
Coking	Adds resid destruction to run medium/sour crude oil

Increasing Complexity ▼

Figure 1. Generic Crude Oil Distillation Column.

U.S. refineries have the capacity to convert some 2.5 Mb/d of petroleum resid into higher value motor fuels. In other words, refineries increased motor fuel production without having to increase nameplate capacity by adding cracking and coking units.[2] Several Midwest refineries have recently added coking/conversion capacity to take advantage of the increasing supply of heavier crude oils from Canada's oil sands projects.

The increasingly available light sweet crude oils produced from unconventional shale resources, like North Dakota's Bakken formation and the Texas Eagle Ford formation, are changing U.S. refining prospects and have revitalized some refinery operations that formerly depended on imported light crude oil. Some have raised questions and concerns about the role of condensates in the market. The domestic market for condensate is somewhat limited, however, and "lease" condensate (a term applied to Outer Continental Shelf production) is subject to current export restrictions.[3]

Processing Condensates in the Oil Field

Condensates are volatile light hydrocarbons in the range of ethane, propane, butane, iso-butane, and iso-pentane (natural gasoline) associated with crude oil.[4] The elevated pressures in oil and gas reservoirs keep condensates in a condensed liquid-state. However, the drop in pressure at the surface can cause condensates to "flash" or vaporize when transferred to the stock tanks. To avoid flashing, the crude oil stream is directed through a separator or "splitter" that separates the condensates into various product streams. A splitter operates as a series of stages, each calibrated to release a particular condensate product based on its characteristic temperature and pressure.[5] The process has similarities to refining, but at temperatures in the range of 100°F to 250°F, and under controlled pressures. In the generic three-stage splitter shown in *Figure 2*, progressively heavier volatiles are separated with each stage ending with degassed crude oil entering the stock tank.

Small Refineries: How They Are Defined and What They Do

From the small business perspective, several legislated provisions define small refineries, but inconsistently. According to the U.S. Small Business Administration (SBA), only refiners that employ fewer than 1,500 employees and produce less than 125,000 barrels per calendar day may qualify for a federal small business contract.[6] By CRS count, 27 refineries make up 50% of U.S. refining capacity. Almost half of the U.S. refineries (some 50 or more) produce less than 125,000 b/d, but in total they make up only 21% of total refining capacity (roughly 18+ Mb/d).

Both Title XV (Ethanol and Motor Fuels) of the Energy Policy Act 2005[7] and Title II (Energy Security Through Increased Production of Biofuels) of the Energy Independence and Security Act of 2007[8] defined a small refinery as having less than 75,000 b/d in average aggregate daily crude oil throughput.

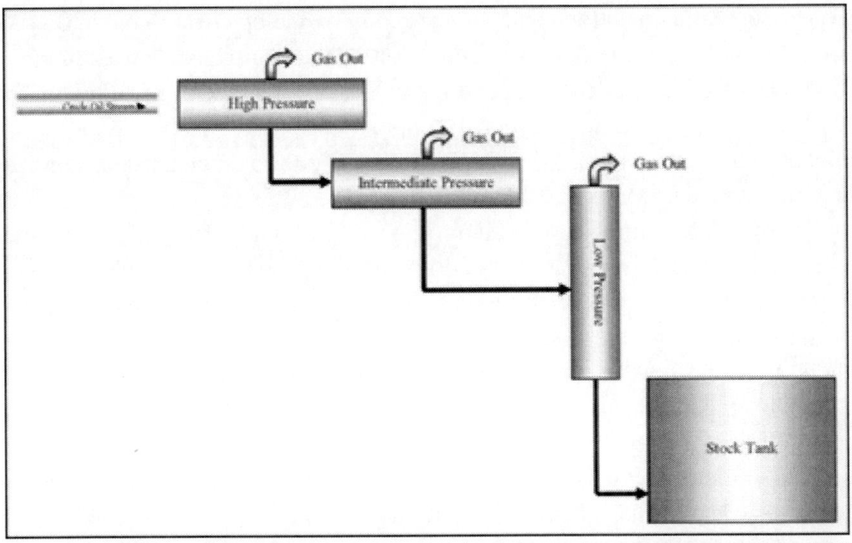

Source: Prepared by CRS from Surface Production Operations, Vol. 1, Design of Oil-Handling Systems and Facilities.

Notes: Crude oil is flashed at an initial pressure and then flashed at successively lower pressures two times before entering stock tank. Condensates flashed will depend on the crude oil, and the operating temperature and pressures of the splitter.

Condensate	Formula	Use
Methane	CH_4	Heating, transportation fuel
Ethane	C_2H_6	Petrochemical feedstock
Propane	C_3H_6	LP Gas heating/transportation fuel
Butane	C_4H_{10}	LP Gas heating/transportation fuel
Iso-butane	C_4H_{10}	Propellant, solvent, refrigerant
Iso-pentane	C_5H_{12}	Natural gasoline/blending

Figure 2. Three Stage Separator.

EPA regulations temporarily exempted small refineries and small volume refineries, which have a capacity below 155,000 b/d, from complying with the Renewable Fuel Standard (RFS) through 2010.[9] The Clean Air Act regulations set 65,000 b/d as the small business threshold for complying with vapor recovery provisions.[10]

Finally, the North American Industry Classification Codes (NAICC) identifies the construction costs for small oil and gas structures at below $33.5 million. However, the smallest refinery currently under construction, a 20,000 b/d refinery in North Dakota, reportedly will cost $350 million.[11] Smaller oilfield equipment used to treat crude oil before it reaches a refinery (e.g., oil and gas separators, splitters, and stabilizers) may line up close to NAICC costs given the mobile and temporary application of the equipment. For example, a "skid-mounted" 2,500 b/d crude oil condensate stabilizer operating in the Texas Eagle Ford shale play cost roughly $400,000 (exclusive of tanks, pipelines, and pumps) in 2012.[12] A permanent 60,000 b/d condensate stabilizer plant might cost upwards of $80 million.

Small business opportunities in refining may strongly depend on a number of factors, including regional demand for refined products, available crude oil supplies, regional price differences, and unique state environmental regulations, among others. In addition, some fuels are easier to produce than others are. The unique "boutique" fuel requirements for some regions can affect the decisions of individual companies to produce, distribute, and market them. While evaluating opportunities would exceed the scope of this report, some key factors have emerged as of late. The frequency that refinery-ownership changes hands suggests that buying an existing refinery offers the most straightforward opportunity to expand capacity quickly. This has been the case for refineries serving regional markets as well as those operating in the major refining centers of the Gulf Coast and the West Coast. The availability of light sweet crude oil in the mid-continent region (North Dakota, Colorado) presents some opportunities for constructing a small topping or hydroskimming refinery. A greater opportunity may lie in installing condensate splitter and stabilizer plants to handle the excess condensate production in North Dakota now being flared,[13] or in Texas. Hess Corp. has applied an even simpler "heater treater" to strip out the light-end hydrocarbons by heating oil to between 80 and 120°F for the oil production from the North Dakota Bakken formation.[14]

As many as 20 other refining projects (with a total capacity of more than 900,000 b/d) are proposed or are in various stages of development.[15] These projects range from splitting and stabilization plants for processing crude oil condensate to "teapot" or mini-refineries. It is uncertain how many projects may actually be completed.

ECONOMIC CHALLENGES

Oil companies can meet increasing/decreasing demand for petroleum products, such as gasoline, in three basic ways. They can build or idle refineries, expand or reduce capacity or capacity utilization rates of existing refineries, or they might choose to export or import petroleum products. These three alternatives allow the industry considerable flexibility in meeting changing short-term demand requirements while still allowing for the implementation of their long-term strategy. Within this framework, any individual refining company can either buy or sell a refinery to change its market position; however, this alternative is not available to the industry as a whole. In general, the industry will choose how to meet demand requirements based on which strategy is the most profitable.

No new complex refineries have been constructed in the United States since the Grayville, LA, facility that opened in 1977. Over 150 refineries have closed since 1982, representing some 1.6 Mb/d in lost refining capacity. However, capacity expansion in existing refineries countered the loss, bringing it up to the current 18+ Mb/d capacity—a 23% increase since the mid-1980s.[16] Between January 2013 and January 2014, refining capacity increased by another 101,000 b/d, the result of investments at existing refineries.

Since 2008, the United States has moved from being a net importer of 1.36 Mb/d of petroleum products to being a net exporter of 1.39 Mb/d in 2013. This shift from net importer to net exporter in less than five years suggests that the United States does not suffer from a shortage of refining capacity. While domestic demand was met, capacity existed to supply foreign markets.

These data suggest that the refining industry has chosen to reduce the number of refineries, increase the capacity of existing refineries, and use foreign markets to meet domestic demands for various petroleum products when it is high and to optimize capacity utilization rates when domestic demand is reduced.

Refining Profitability

Over the period 2006 through 2009, as shown in *Figure 1*, refiners' profits generally declined. This may be attributed to several factors, including high oil prices that peaked in 2008, coupled with weak demand due to the recession that began in December of 2007. *Figure 1* also shows that North American refiners' profits increased from 2009 through 2012, because of recovering demand and discounted crude oil supplies that resulted from expanded production from U.S. tight oil deposits. Profits decreased, and then turned upward, from 2012 through the first quarter of 2014. The profit volatility over time shown in *Figure 3* is common for the refining industry.

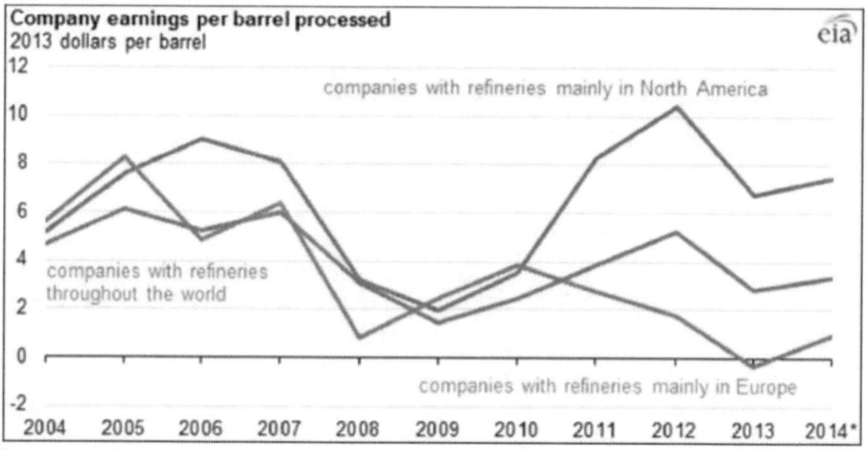

Source: U.S Energy Information Administration; http://www.eia.gov/todayinenergy/detail.cfm?id=16571.

Notes: * 2014 estimated.

Figure 3. North American Refinery Profitability.

Uncertainty with respect to industry profitability is an important reason why the decision to build a new refinery in the United States has not been taken. Desirable multi-billion dollar investment projects should offer companies predictable profits in the end, reflecting the long-run time frame for refinery investment and operation. Generally, North American refinery earnings per barrel processed have been positive.

Refining Capital Investment

The refining industry undertakes capital investment for a variety of reasons, including expanding existing or creating new production facilities, implementing new or enhanced technology, regulatory compliance, and adapting existing refineries to available crude oil streams. With the exception of investments to insure compliance with environmental and other regulations, other activities may be considered to be competing for capital resources. When profitability declines or regulations tighten, the resource pool available for capital investment is reduced. *Table 1* shows the capital budget expenditures of the refining industry.

U.S. refining capital investment peaked in 2008, reflecting rapidly increasing world demand for petroleum products. A 22% decline in capital investment from 2008 through 2009 was followed by a further 50% decline from 2009 through 2010. More recently, the surge of petroleum product exports, coupled with investments to take advantage of new supplies of discounted light, sweet crude oil have seen investments increase. An example of the importance of new light, sweet crude December 2014. This new refinery, with a capacity of 20,000 b/d, is a direct result of the expansion of oil production in the Bakken fields. In addition, it has been reported that as many as five additional small refineries might be constructed in North Dakota.[17]

Table 1. U.S. Refining Industry Capital Budget Expenditures 2005-2013

Year	$ Billion
2005	7.2
2006	9.0
2007	8.3
2008	13.0
2009	10.1
2010	5.3
2011	9.2
2012	13.0
2013*	12.9

Source: Oil and Gas Journal, March 7, 2011; Oil and Gas Journal, March 4, 2013.
Notes: * 2013 Estimated.

One part of the capital investment decision is the time and cost of permitting a new site. Many observers have raised this issue with respect to the lack of refinery construction in the United States. One reported example of permitting issues purportedly associated with refinery construction delays is the Arizona Clean Fuels Refinery, which planned to refine Mexican crude oil in Arizona for use in the western United States.[18] Interest in constructing this approximately $3 billion facility began in 1998 and took seven years and reportedly millions of dollars of cost for permitting, and has yet to be constructed. However, the project has also experienced financing difficulties, local zoning changes, and legal challenges.[19]

For capital investment in new refineries to be desirable, certain market conditions should prevail. An important prerequisite is an expectation of growth in demand, especially in the transportation fuels, gasoline and distillates. The availability of stable, affordable crude oil supplies, and certainty with respect to the regulatory environment, are also important.

While growing demand for petroleum products is a key element in refining profitability, the U.S. demand for gasoline declined by about 5% from 2007 through 2013. Although this fall in demand was likely tied to the recession, and might be considered temporary, other factors may continue to reduce demand, or at least demand growth. Among these factors are rising Corporate Average Fuel Economy (CAFE) Standards, which are set to increase to roughly 50 miles per gallon by 2025. Full implementation of these standards could reduce the U.S. demand for motor fuel by 12 Mb/d, as discussed next.[20] These savings represent oil that the domestic refining industry will not be required to process over the next decade, reducing the need for new capacity.

The growth in the use of bio-based ethanol as a component of gasoline has also contributed to reduced demand for crude oil per gallon of gasoline. To a lesser extent, this is true for biodiesel and diesel fuel as well. In addition, the potential growth in alternative fuel vehicles, including hybrid and electric vehicles and natural gas-fueled vehicles, could further demand growth.

While the growing U.S. crude oil and natural gas production from unconventional resources would seem to assure the refining industry that a reliable source of domestic feedstock would be available, several problems have developed. The quality of very light sweet crude oil supplies is in one sense very high. However, larger and more complex refineries have positioned themselves to take advantage of lower quality heavy sour crude. For them, light sweet crude represents an opportunity that will depend on a competitive price advantage it offers in refining. Having made considerable investments in crack and coke heavy feedstock, these refiners now face the decision to bypass their conversion capacity and shift to light crude. This processing re-orientation spells increased competition for smaller refineries that had forgone investment in upgrades and continued to process diminishing light crude supplies.

Refinery and fuel environmental regulations are covered in the next section of this report. However, it can be noted that given any level of refining industry capital investment, tightening regulations may impute a cost to any refiner that wishes to remain in operation.

Vehicle and Fuel Regulatory Requirements

Two key vehicle and fuel regulatory standards—the coordinated Corporate Average Fuel Economy (CAFE)/Green House Gas (GHG) vehicle emission standards and the Renewable Fuel Standard (RFS)—have likely affected the refining industry by reducing the consumption of petroleum fuels.[21] Compliance with the RFS may also increase the costs for some refiners, as well as present other challenges.

Corporate Average Fuel Economy (CAFE)/GHG Vehicle Emission Standards

The federal government has regulated vehicle fuel economy since 1978 and, more recently, aligned these efforts with new GHG vehicle emission standards in a joint national program aimed at reducing petroleum consumption and GHG emissions from the transportation sector.[22] As required by the Energy Policy and Conservation Act of 1975 (EPCA),[23] the National Highway Traffic Safety Administration (NHTSA) sets Corporate Average Fuel Economy (CAFE) standards for passenger cars and light trucks. These standards were raised under the Energy Independence and Security Act of 2007 (EISA),[24] and, subsequently, coupled with vehicle GHG standards as administered by the Environmental Protection Agency (EPA) through its authority under the Clean Air Act and subsequent amendments.[25]

Fuel consumption and greenhouse gas (GHG) emissions from motor vehicles are closely linked. The vast majority of vehicle GHG emissions result from the burning of petroleum products, so reducing vehicle fuel consumption is the most direct means of reducing emissions. For these reasons, the Obama Administration has issued joint rules on vehicle fuel economy and GHG emissions for model year (MY) 2012-2016 passenger cars and light trucks,[26] MY2014-MY2018 medium- and heavy-duty trucks,[27] and MY2017-MY2025 passenger cars and light trucks.[28] In addition, EPA and NHTSA have begun drafting a second phase of fuel economy/GHG standards for FY2019 and later medium- and heavy-duty trucks; proposed standards are expected in March 2015.

NHTSA and EPA expect that combined new passenger car and light truck CAFE standards will rise to as much as 41.0 miles per gallon (mpg) in MY2021 and 49.7 mpg in MY2025, up from 34.1 mpg in MY2016. EPA and NHTSA expect that the combined National Program for MYs 2012-2016 and MYs 2017-2025 is projected to reduce U.S. oil use by more than 2 Mb/d in 2025.[29] Critics have challenged the Administration's assumptions, countering that the costs will be higher and could lead to a drop in new vehicle sales.

Though the standards do not require changes at the refinery level, they can affect refineries indirectly by contributing to improvements in the overall efficiency of the vehicle fleet and, therefore, reducing fuel consumption.[30] These standards will certainly reduce petroleum consumption in the long term, but short-term effects on fuel consumption and thus the refining industry are less clear. For example, DOE officials stated that thus far "the impact of the standards has been limited because they affect new car sales, and there are a relatively small number of new vehicles in the overall fleet."[31] However, other stakeholders reported to GAO that with estimated fuel savings of such magnitude, "the CAFE and GHG vehicle emissions standards have likely had a relatively large impact on petroleum demand declines in the past few years."[32]

According to the U.S. Government Accountability Office, "CAFE and GHG vehicle emission standards have contributed to reductions in the consumption of petroleum fuels, but the extent is unclear."[33]

How new and existing refinery operations—both large and small—adjust in the short and long terms to the impact of reduced domestic demand for petroleum transportation fuels will be dependent upon their respective market projections.

Renewable Fuel Standard (RFS)

The Renewable Fuel Standard (RFS) requires the use of biofuels in transportation fuel.[34] The RFS was established in the Energy Policy Act of 2005 (P.L. 109-58) and significantly expanded in the Energy Independence and Security Act of 2007 (P.L. 110-140). The vast majority of the mandate is met using corn-based ethanol blended into gasoline (although there is no explicit mandate to use ethanol or that it be blended). For 2013, the RFS mandated the use of 16.55 billion gallons of biofuels, of which roughly 13.3 billion gallons came from corn ethanol.[35] At the 2013 level, renewable fuels replace roughly 10% of overall demand for motor vehicle fuels, and displace roughly 1Mb/d of gasoline and diesel consumption. EISA requires a further increase in the use of renewables, to 36 billion gallons by 2022, further reducing demand for petroleum-based fuel.

Renewable Identification Numbers (RINs) are the compliance mechanism for the RFS.[36] After each year, a refiner or other obligated party must submit to EPA a sufficient number of RINs to cover its obligation (which is prorated for each obligated party based on gasoline and diesel fuel produced/supplied in that year). RINs are generated by biofuel producers and are sold with the fuel. Biofuel purchasers—generally gasoline blenders—typically "separate" the

RINs from the physical fuel when they receive and blend the ethanol into gasoline. Once a RIN has been separated the owner may sell the RIN to another party, be that another obligated entity in need of additional RINs or a third-party trader.

In the case when an independent refiner is not integrated with downstream operations, an obligated party may only be able to secure RINs through this secondary market. If supplies of RINs are tight, then these parties may need to pay market prices for RINs, which could reflect the marginal cost of compliance as opposed to the industry average. If RINs are trading at a high enough price, non-integrated refiners could face higher compliance costs than their competitors. Thus, although there is no direct effect on the construction of new refineries from the RFS, the need to secure RINs may present an economic hurdle to a new, independent refinery.

Because of concerns over compliance costs, P.L. 109-58 gave small volume refineries (75,000 bpd or less) a temporary exemption from the RFS requirements through the end of 2010. EPA extended this exemption to small refiners (SBA definition, see "Tier 3 Sulfur Standards" below). Later, Congress directed the DOE to study whether the RFS would lead to differential economic hardship on some or all small refiners and that EPA should extend the exemption for those refineries. DOE found that some small refiners would face disproportionate economic hardship and that those refiners should be exempted.[37] For 2011 and 2012, EPA exempted 13 small refineries (with their names redacted to protect confidential business information).[38] EPA approved a single exemption for 2013,[39] and proposed no exemptions for 2014.[40]

At the state level, the California legislature passed, and then-Governor Schwarzenegger signed, a 2007 law requiring statewide reductions in GHG emissions by 2020. Among other regulations to implement the law, the California Air Resources Board established a Low-Carbon Fuel Standard (LCFS) requiring refiners to reduce the carbon intensity of the fuels they provide to the California market. By 2020, the regulations require a roughly 10% reduction from 2010 levels. In general, it is expected that most of the requirement will be met using various lower-carbon biofuels, although California's standards for lifecycle emissions are stringent, and some biofuels actually have higher emissions than gasoline or diesel fuel under the rule. The interactions between the California program and the federal Renewable Fuel Standard (RFS) could be complex. It is unclear how much the LCFS will raise refiners' costs, as early reviews of the program have shown little effect on the market and compliance credits are currently trading at low levels. However, as the program becomes more stringent, the compliance costs are likely to increase.

Tier 3 Sulfur Standards

In April 2014, EPA finalized new gasoline sulfur standards.[41] Beginning in 2017 the average sulfur content in most gasoline will be reduced from 30 parts per million (ppm) to 10 ppm. This is part of a larger strategy to reduce emissions from both new and existing passenger vehicles. The sulfur standards were issued along with new "Tier 3" emissions standards for new vehicles.[42] The lower sulfur level limits pollutant formation in existing vehicles and allows the use of more advanced emissions systems in new vehicles—systems that are highly sensitive to the amount of sulfur in the fuel.

In general, the cost of refining gasoline is expected to increase, as investments must be made in desulfurization technology. EPA estimates that the cost of compliance will average below one cent per gallon, with some refiners facing two to three cents per gallon, although industry stakeholders contend that the price could be significantly higher—as much as six to nine cents per gallon for some refiners.[43] The studies generally found that refiners would expand existing desulfurization units instead of constructing new equipment, although it does not appear that they looked at the effects on the construction of completely new refineries. Again, the differential impacts of these standards may be more or less likely to affect new or small refinery vis-à-vis their competitors.

In general, small refiners are given more flexibility in compliance with fuel standards compared to larger refiners. For most fuel standards, EPA has two classes of small refiners:

- Small refiners: a refiner with 1,500 or fewer employees (Small Business Administration (SBA) definition) and annual crude oil capacity of 155,000 b/d or less; or
- Small volume refineries: a refiner with 75,000 b/d of crude capacity, regardless of the number of employees

For the Tier 3 standards, small refiners and small volume refineries will be granted an additional three years to come in compliance with the standards—until 2020—although they would be able to generate tradable credits for early compliance with the standards.

ENVIRONMENTAL PERMITTING AND REGULATORY REQUIREMENTS

As with any industrial facility involving potential air emissions, water use, and waste generation, certain environmental permitting and regulatory requirements apply to the construction and operation of refineries. The specific requirements, and the degree of difficulty or ease in meeting them, will depend not only on the nature of the project, but perhaps more importantly on the location of the project. Congress included in the Energy Policy Act of 2005 (EPAct 2005)[44] a provisions on "Refinery Revitalization" intended to improve permitting coordination between EPA and various federal and state agencies.

In general, state or local agencies handle environmental permitting and other regulatory operations, in part because EPA has delegated authority to them for many federal environmental programs.[45] Accordingly, state and local agencies largely implement permitting, inspections, monitoring, and enforcement, and, in some cases, standards setting. In addition, many states have enacted statutes and promulgated certain requirements independent of EPA to address specific issues within their states.

While air pollution permitting and standards are typically the biggest regulatory challenge for new refineries, the specific location and situation of a proposed new refinery may pose other regulatory challenges (e.g., cooling water supply may be limited, particularly in arid areas). This report focuses on air pollution permitting and standards for new and modified refineries, as well as some additional considerations such as the requirements on the fuels produced by the refineries.

Air Pollution Permitting

Under the Clean Air Act (CAA),[46] EPA is authorized to take steps to address air emissions from stationary sources, including refineries. The agency has regulated emissions of certain air pollutants from new and modified stationary sources for several decades under the authority of various sections of the CAA, including air permitting requirements, Section 111 New Source Performance Standards (NSPS), and Section 112 National Emission Standards for Hazardous Air Pollutants (NESHAPS).

While air permitting of major sources of pollution is a requirement under the CAA, all 50 states have delegated responsibility for permitting. The Clean

Air Act amendments of 1990 (P.L. 101- 549, 42 U.S.C. §§7470-7479 and §§7661-7661f) require that major industrial sources of air pollutants obtain both construction and operating permits. These permits are intended to enhance compliance by detailing for each covered facility all the emission control requirements to which the facility is subject.

A Prevention of Significant Deterioration (PSD) permit applies to the construction of a new or modified facility that will have "major" and "significant" amounts of air pollution for any criteria pollutant.[47] A PSD permit is required before a new source is constructed, or before changes or modifications are made at an existing source of air pollution. Either EPA or the designated state permitting authority may issue the permit (which must specify the construction allowed, the emission limits, and frequency of operation for the equipment being permitted). PSD requires

1. installation of the Best Available Control Technology,[48]
2. an air quality analysis,
3. an additional impacts analysis, and
4. public participation.

New and modified petroleum refineries are listed as a source category for PSD requirements.

A Title V operating permit effectively consolidates and coordinates the often-complex permitting requirements for the various emissions sources characteristic of refinery operations.[49] Emissions from a refinery arise from various specific operations, including cracker towers, coker units, gas combustion turbines, boilers, and storage tanks. The emissions from these operations typically include criteria pollutants such as volatile organic compounds (VOCs), nitrogen oxides (NO_x), sulfur oxides (SO_x), carbon monoxide (CO), and particulate matter (PM or "soot"), as well as various hazardous air pollutants (HAPs, e.g., benzene, toluene and naphthalene). Title V permits generally address "major sources" of air pollution, defined in the CAA as stationary facilities that emit or have the potential to emit 100 tons or more per year of any regulated pollutant or combination of pollutants. Title V also covers the following:

1. sources in nonattainment areas (described below) that emit as little as 10 tons per year of VOCs, depending on the region's nonattainment status;

2. sources subject to New Source Performance Standards (NSPS, described below);
3. regulated sources of HAP emissions (any source that emits more than 10 tons per year of an individual HAP or more than 25 tons per year of any combination);
4. sources required to have new source or modification permits under Title I of the act.

A federal Title V permit would likely be required for an "average" (approximately 140,000 BPD) new oil refinery, but may depend on the details of the State Implementation Plan where the refinery is to be located.

Specific permit requirements and conditions depend on where a potential source is located. Air pollution control requirements will be more stringent for a refinery located in a "nonattainment" area—an area where air quality does not meet the minimum National Ambient Air Quality Standards (NAAQS)[50] established to address the health impacts from exposure to the six criteria air pollutants. For example, a refinery seeking a permit in a nonattainment area (e.g., Houston-Galveston, TX, or Philadelphia, PA) could be required to meet the "Lowest Achievable Emissions Rate"[51] for its processes. In addition, any expected emissions potential must be offset by sufficient Emission Reduction Credits (ERCs), which are usually generated through emissions reductions at existing sources or through the shutdown of existing sources. For example, a refinery seeking a permit in an area classified as "severe nonattainment," may be required to seek ERCs at a rate of 1.3:1 compared to its expected emissions potential.

The permitting requirements for a small refinery will depend on the technical specifications of what is being permitted and where. The applicability of permitting requirements is based on the "potential to emit" from facilities on contiguous or adjacent properties and under the common control of a person or entity. Thus, certain operations—depending upon their size, status, and "potential to emit"—may not require a Title V permit and could be permitted under other state permitting regulations (e.g., synthetic minor permit). The number and kinds of permitting required by a small refinery and its specific equipment will rest on these case-specific determinations.

While these air permitting issues pertain to any facility, not solely refineries, with a similar potential to emit, they are summarized here because they are often viewed as a limitation to the construction of a new refinery or the modification of an old one. In some cases, the availability of offsets may pose a constraint to refinery permitting in certain nonattainment areas. In addition, the often-complex calculations may require a significant investment in technical expertise by both the facility operator seeking a permit and the permitting agency, typically at the state or local level.

Air Pollution Standards

One of the significant air pollution regulatory requirements for a new or modified refinery includes industry-specific New Source Performance Standards (NSPS) and National Emission Standards for Hazardous Air Pollutants (NESHAPS).

Industry-specific NSPS requirements have been promulgated by EPA to implement Section 111(b) of the CAA and are issued for categories of sources that cause, or contribute significantly to, air pollution, which may reasonably be anticipated to endanger public health or welfare. The existing NSPS for petroleum refineries address PM, CO, and SO_2 emissions.[52] The standards include emissions limitations and work practice standards for fluid catalytic cracking units, fluid coking units, delayed coking units, fuel gas-combustion devices, and sulfur recovery plants, as well as requirement for monitoring and recordkeeping, among other things. An NSPS sets an emission rate limit on a given piece of equipment or activity that is equivalent to the "best system of emission reductions" that is currently and adequately demonstrated by industry practice, taking into account cost and other factors. NSPS must be implemented by all refineries, regardless of size.

Industry-specific NESHAPS requirements have been promulgated by EPA to implement Section 112 of the CAA and are issued for categories of new, reconstructed, or existing sources that emit HAPs as defined under Section 112(b) of the CAA.[53] EPA has promulgated National Emission Standards for Petroleum Refineries, including National Uniform Emission Standards for Heat Exchange Systems (Refinery MACT 1) and Catalytic Cracking, Catalytic Reforming, and Sulfur Plant Units (Refinery MACT 2).[54] The rules establish emission limits for HAPs from petroleum refineries and requirements to demonstrate initial and continuous compliance with the emission limitations and work practice standards. On May 15, 2014, EPA issued a proposed rule that would further control HAPs from petroleum refineries.[55] The rule proposes additional emission control requirements for storage tanks, flares, and coking units at petroleum refineries as well as the monitoring of air concentrations at the fence-line of refinery facilities to ensure proposed standards are being met and that neighboring communities are not being exposed to unintended emissions. As with the NSPS, the refinery NESHAPS must be implemented by all refineries, regardless of size, to the extent they are categorized as either a major or area source of HAPs.

Starting in 2011, new refinery construction, refinery expansion, or modification of certain existing refineries became subject to the use of the "best available control technology" for greenhouse gas (GHG) emissions. The best available control technology is determined for each facility based on an analysis of available technologies considering cost and other factors. According to EPA, in most cases, the best available control technology selected for GHGs is energy efficiency improvements. For example, for refineries, this could involve the installation of heat recovery units, which capture and use otherwise wasted heat in the refinery process. Such energy efficiency improvements can lower GHG emissions and other pollutants while reducing fuel consumption and saving money.

Currently, there are no federal GHG emissions control standards for petroleum refineries. If promulgated, however, such control requirements standards could affect the refining industry, depending on the nature and timing of the standards.[56] In December 2010, EPA entered into two proposed settlement agreements to issue rules that will address GHG emissions from fossil fuel-fired power plants and refineries.[57] In the settlement, EPA agreed to promulgate by November 10, 2012, NSPS for GHG emissions from new and modified petroleum refineries, as well as guidelines for existing petroleum refineries. However, this deadline has since passed and it is unclear when EPA will issue a proposal or a final rule or what it would look like. Although it is unclear when, or if, federal regulations would be proposed, California's requirements under A.B.32 are expected to affect refineries there and other states may follow.

CONCLUSION

The United States is experiencing an energy renaissance in producing light crude oil and natural gas. Unforeseen just a few years ago, these new resources present some challenges as well as new opportunities to U.S refiners. Rather than U.S. refiners having to chase import suppliers, as has been the past case, some U.S. producers now find foreign brokers chasing them.

The refining industry has largely chosen to reduce the number of refineries, increase the capacity of existing refineries rather than construct new ones, and utilize foreign markets to meet domestic demands for various petroleum products when demand is high and to optimize capacity utilization rates when domestic demand is reduced. Given the significant expansions of refineries in the industry consolidations during the past decades, there appears to be little general evidence that such growth in refinery capacity has been impeded significantly by state and federal environmental permitting requirements. Complex refineries that locked into imported heavy crude oils to run through their cracking and coking plants are taking the new light crudes and adapting their refining economics to run them. The shift could spell more competition for smaller refineries that had depended on light crudes and that returned to profitability with the crudes' increased availability. The possibility of bypassing conventional refining altogether and exporting light crude (condensate) directly is a newly emerging development. The latter case presents possible opportunities for small businesses to enter the petroleum supply sector upstream of refining.

In light of the restrictions on exporting U.S.-produced crude oil and condensate, Congress may consider whether the definition of a refinery includes similar processes that produce product streams based on their physical properties. Congress may also wish to assess whether small business enterprises face unique challenges to entering refining, or unique opportunities now that the United States has become a net exporter of refined petroleum products.

End Notes

[1] CRS Report R41478, The U.S. Oil Refining Industry: Background in Changing Markets and Fuel Policies, by Anthony Andrews et al.

[2] Nameplate capacity refers to the throughput capacity in barrels per day of a refinery's atmospheric distillation tower.

[3] Christian Berthelsen and Lynn Cook, "U.S. Set to Export First Oil Since '70s," Wall Street Journal, Eastern edition (New York, NY), June 25, 2014; and "Energy Industry Buzzes at Oil Export Opening," Wall Street Journal, July 15, 2014. For more information on oil exports, see CRS Report R43442, U.S. Crude Oil Export Policy: Background and Considerations, by Phillip Brown et al.

[4] Propane and butane are also referred to as liquefied petroleum gas (LPG).

[5] Ken Arnold and Maurice Stewart, Surface Production Operations, Volume 1, Design of Oil-Handling Systems and Facilities (Houston, Texas: Gulf Publishing Company, 1995), p. 33.

[6] "NAICS code 324110—For purposes of Government procurement, the petroleum refiner must be a concern that has neither more than 1,500 employees nor more than 125,000 barrels per calendar day total Operable Atmospheric Crude Oil Distillation capacity. Capacity includes owned or leased facilities as well as facilities under a processing agreement or an arrangement such as an exchange agreement or a throughput. The total product to be

delivered under the contract must be at least 90 percent refined by the successful bidder from either crude oil or bona fide feedstocks." U.S. Small Business Administration, "Table of Small Business Size Standards Matched to North American Industry Classification System Codes," http://www.sba.gov/sites/default/files/files/Size_Standards_Table.pdf.

[7] P.L. 109-58, and codified in 40 C.F.R. 80.1101 (g). "Small refinery, for this subpart only, means a refinery for which the average aggregate daily crude oil throughput for calendar year 2006 (as determined by dividing the aggregate throughput for the calendar year by the number of days in the calendar year) does not exceed 75,000 barrels."

[8] P.L. 110-140.

[9] 40 C.F.R. 80.1142—What Are the Provisions for Small Refiners Under the RFS Program?

[10] 42 U.S.C. 7625—Vapor Recovery for Small Business Marketers of Petroleum Products.

[11] Rachael Seeley, "MDU Expands North Dakota Presence with Greenfield Refinery," Unconventional Oil and Gas Report, May/June 2014.

[12] Personal communication with Bill Bowers, Valerus Inc., July 8, 2014. See "Eagle Ford Complete Well Pad Solution," http://www.valerus.com/knowledge-center/.

[13] Russell Gold and Chester Dawson, "Firms Leave Oil from North Dakota Fracking Prone to Explosion," Wall Street Journal, July 8, 2014, pp. B1, B2.

[14] Russell Gold and Chester Dawson, "Firms Leave Oil from North Dakota Fracking Prone to Explosion," Wall Street Journal, July 8, 2014, pp. B1, B2.

[15] Christian Bethelsen and Lynn Cook, "U.S. Set to Export First Oil Since '70s," Wall Street Journal, June 25, 2014.

[16] 14.4 Mb/d in 1985 vs. 17.7 Mb/d in 2014; see U.S. Energy Information Administration, Number and Capacity of Petroleum Refineries, http://www.eia.gov/dnav/pet/ pet_pnp_cap1_dcu_nus_a.htm.

[17] Oil Daily, July 3, 2014, pg.1.

[18] PADD: Petroleum Administration for Defense District. PADD 5 includes Alaska, Washington, Oregon, California, Nevada, Arizona, and Hawaii.

[19] The refinery plans appear to be on hold, with no press releases since March 2009.

[20] The White House, "Obama Administration Finalizes Historic 54.5 MPG Fuel Efficiency Standards," press release, August 28, 2012, http://www.whitehouse.gov/the-press-office/2012/08/28/obama-administration-finalizes-historic-545-mpg-fuel-efficiency-standard.

[21] EPA Tier 3 sulfur standards are discussed in the section "Environmental Permitting and Regulatory Requirements."

[22] For a more detailed discussion on fuel economy standards, see CRS Report R42721, Automobile and Truck Fuel Economy (CAFE) and Greenhouse Gas Standards, by Brent D. Yacobucci, Bill Canis, and Richard K. Lattanzio.

[23] P.L. 94-163.

[24] P.L. 110-140

[25] 42 U.S.C. 7521 et seq.

[26] Environmental Protection Agency and National Highway Traffic Safety Administration, "Light-Duty Vehicle Greenhouse Gas Emission Standards and Corporate Average Fuel Economy Standards; Final Rule," 75 Federal Register 25324, May 7, 2010.

[27] Environmental Protection Agency and National Highway Traffic Safety Administration, "Greenhouse Gas Emissions Standards and Fuel Efficiency Standards for Medium- and Heavy-Duty Engines and Vehicles; Final Rule," 76 Federal Register 57106, September 15, 2011.

[28] The CAFE standards only apply through MY2021 because of stipulations in the fuel economy law. NHTSA will need to issue additional regulations for MY2022 onward, while EPA has the authority to set GHG standards for MY2025 and beyond. EPA and NHTSA, "2017 and Later Model Year Light-Duty Vehicle Greenhouse Gas Emissions and Corporate Average Fuel Economy Standards," 77 Federal Register 62623, December 14, 2012.

[29] EPA, "EPA and NHTSA Set Standards to Reduce Greenhouse Gases and Improve Fuel Economy for Model Years 2017-2025 Cars and Light Trucks," Regulatory Announcement, August 2012.

[30] U.S. Government Accountability Office, Petroleum Refining: Industry's Outlook Depends on Market Changes and Key Environmental Regulations, GAO-14-249, March 2014, p. 26.

[31] U.S. Government Accountability Office, Petroleum Refining: Industry's Outlook Depends on Market Changes and Key Environmental Regulations, GAO-14-249, March 2014, p. 27.

[32] Ibid.

[33] U.S. Government Accountability Office, Petroleum Refining: Industry's Outlook Depends on Market Changes and Key Environmental Regulations, GAO-14-249, March 2014.

[34] For more information, see CRS Report R40155, Renewable Fuel Standard (RFS): Overview and Issues, by Randy Schnepf and Brent D. Yacobucci.

[35] Environmental Protection Agency, 2013 RFS Data, June 7, 2014, http://www.epa.gov/otaq/fuels/rfsdata/ 2013emts.htm.

[36] For more information, see CRS Report R42824, Analysis of Renewable Identification Numbers (RINs) in the Renewable Fuel Standard (RFS), by Brent D. Yacobucci.

[37] Department of Energy, Small Refinery Exemption Study: An Investigation into Disproportionate Economic Hardship, March 2011.

[38] Environmental Protection Agency, Small Refinery Exemptions Under the Renewable Fuel Standard Program, May 2011.

[39] Environmental Protection Agency, "Regulation of Fuels and Fuel Additives: 2013 Renewable Fuel Standards; Final Rule," 78 Federal Register 49798, August 15, 2013, https://federalregister.gov/a/2013-02794.

[40] Environmental Protection Agency, "2014 Standards for the Renewable Fuel Standard Program; Proposed Rule," 78 Federal Register 71732, November 29, 2013, https://federalregister.gov/a/2013-28155.

[41] Environmental Protection Agency, "Control of Air Pollution from Motor Vehicles: Tier 3 Motor Vehicle Emission and Fuel Standards; Final Rule," 79 Federal Register 23618, April 28, 2014, https://federalregister.gov/a/2014-06954.

[42] For more information, see CRS Report R43497, Tier 3 Motor Vehicle Emission and Fuel Standards, by Richard K. Lattanzio and James E. McCarthy.

[43] David C. Tamm and Kevin P. Milburn, Addendum to Potential Supply and Cost Impacts of Lower Sulfur, Lower RVP Gasoline, Baker & O'Brien Incorporated, March 2012, p. 12, http://www.api.org/news-and-media/news/newsitems/2013/march-2013/~/media/Files/News/2012/12-March/Addendum-Potential-Impacts-of-Lower-Sulfur-Lower-RVPGasoline-Report.pdf.

[44] P.L. 109-58

[45] See delegation of authority by environmental act at http://www.ecos.org/section/states/enviro_actlist. Note that several state environmental agencies, including Ohio, Illinois, and California, are also called "Environmental Protection Agency," so distinguishing between federal and state regulatory agencies is important.

[46] 42 U.S.C. 7401 et seq. For a summary of the CAA and EPA's air and radiation activities and its authorities, see EPA's website at http://www.epa.gov/air/basic.html; and CRS Report

RL30853, Clean Air Act: A Summary of the Act and Its Major Requirements, by James E. McCarthy and Claudia Copeland.

[47] "Major" is defined as emitting or having the potential to emit 100 tons per year (tpy) or more of any criteria pollutant. "Significant" refers to thresholds assigned to each criteria pollutant and certain non-criteria pollutants (e.g., the significance threshold is 40 tpy for NOx and 15 tpy for PM-10). "Criteria" pollutants include carbon monoxide, lead, nitrogen dioxide, ozone, particulate matter, and sulfur dioxide.

[48] Best Available Control Technology (BACT) is an emissions limitation which is based on the maximum degree of control that can be achieved at a major stationary source. It is a case-by-case decision that considers energy, environmental, and economic impact. BACT can be add-on control equipment or modification of the production processes or methods.

[49] For a full discussion of the Title V requirements, see CRS Report RL33632, Clean Air Permitting: Implementation and Issues, by Claudia Copeland.

[50] The Clean Air Act requires EPA to set National Ambient Air Quality Standards (40 CFR part 50) for pollutants considered harmful to public health and the environment. EPA has set National Ambient Air Quality Standards for six principal pollutants, which are called "criteria" pollutants, and include carbon monoxide, lead, nitrogen dioxide, ozone, particulate matter, and sulfur dioxide. For a list of current standards, see EPA at http://www.epa.gov/air/criteria.html.

[51] Lowest Achievable Emissions Rate (LAER) is the most stringent emission limitation derived from either of the following: (1) the most stringent emission limitation contained in the implementation plan of any State for such class or category of source; or (2) the most stringent emission limitation achieved in practice by such class or category of source.

[52] 40 C.F.R. 60 Subparts J and Ja—Standards of Performance for New Stationary Sources.

[53] Hazardous air pollutants, also known as HAPs, toxic air pollutants or air toxics, are those pollutants that cause or may cause cancer or other serious health effects, such as reproductive effects or birth defects, or adverse environmental and ecological effects. EPA is required to control 187 hazardous air pollutants that are listed in Section 112 (b) of the CAA.

[54] 40 C.F.R. 63 Subpart CC—National Emission Standards for Petroleum Refineries (Includes National Uniform Emission Standards for Heat Exchange Systems) and 40 C.F.R. 63 Subpart UUU—National Emission Standards for Petroleum Refineries (Catalytic Cracking, Catalytic Reforming, and Sulfur Plant Units).

[55] Environmental Protection Agency, "Petroleum Refinery Sector Risk and Technology Review and New Source Performance Standards, Proposed Rule," 79 Federal Register 36879, June 30, 2014, https://federalregister.gov/a/2014- 12167.

[56] Following several states that had previously promulgated mandatory reporting of GHG emissions, EPA promulgated a rule on October 30, 2009, requiring reporting of GHG emissions from most refineries (i.e., sources that in general emit 25,000 metric tons or more of carbon dioxide equivalent per year). See U.S. Environmental Protection Agency, "Mandatory Reporting of Greenhouse Gases; Final Rule," 74 Federal Register 56260, October 30, 2009.

[57] Environmental Protection Agency, "Settlement Agreement for Petroleum Refineries—December 23, 2010," http://www2.epa.gov/sites/production/files/2013-09/documents/refineryghgsettlement.pdf.

In: U.S. Petroleum Refining
Editor: Roger M. Waters

ISBN: 978-1-63463-336-9

Chapter 3

THE U.S. OIL REFINING INDUSTRY: BACKGROUND IN CHANGING MARKETS AND FUEL POLICIES*

Anthony Andrews, Robert Pirog and Molly F. Sherlock

SUMMARY

A decade ago, 158 refineries operated in the United States and its territories and sporadic refinery outages led many policy makers to advocate new refinery construction. Fears that crude oil production was in decline also led to policies promoting alternative fuels and increased vehicle fuel efficiency. Since the summer 2008 peak in crude oil prices, however, the U.S. demand for refined petroleum products has declined, and the outlook for the petroleum refining industry in the United States has changed.

In response to weak demand for gasoline and other refined products, refinery operators have begun cutting back capacity, idling, and, in a few cases, permanently closing their refineries. By current count, 124 refineries now produce fuel in addition to 13 refineries that produce lubricating oils and asphalt. Even as the number of refineries has decreased, operable refining capacity has actually increased over the past decade, from 16.5 million barrels/day to over 18 million barrels/day.

* This is an edited, reformatted and augmented version of a Congressional Research Service publication, R41478, dated November 22, 2010.

Cyclical economic factors aside, U.S. refiners now face the potential of long-term decreased demand for their products. This is the result of legislative and regulatory efforts that were originally intended, in part, to accommodate the growing demand for petroleum products, but which may now displace some of that demand. These efforts include such policies as increasing the volume of ethanol in the gasoline supply, improving vehicle fuel efficiency, and encouraging the purchase of vehicles powered by natural gas or electricity.

Since the Clean Air Act Amendments, 15 distinctly formulated boutique fuels are required in portions of 12 states. H.R. 392, the Boutique Fuel Reduction Act of 2009, would further amend` the Clean Air Act to add temporary waivers for boutique fuels due to unexpected problems with distribution and give EPA authority to reduce the number of boutique fuels. The 2005 Energy Policy Act created the Renewable Fuel Program to substitute increasing volumes of renewable fuel for gasoline. The 2007 Energy Independence and Security Act expanded the program to cover transportation fuels in general, extended the program to calendar year 2022, and increased the target volume to 36 billion gallons renewable fuel annually. The 2008 Food, Conservation and Energy Act of 2008 reduced some of the federal subsidies and tax breaks favoring ethanol production. A 2007U.S. Supreme Court ruling found that EPA has the authority under the Clean Air Act to regulate carbon dioxide (CO2) emissions from automobiles. Though the ruling applied to automobiles, it had wider implications. In response to the FY2008 Consolidated Appropriations Act (H.R. 2764; P.L. 110-161), EPA issued the Mandatory Reporting of Greenhouse Gases Rule that requires suppliers of fossil fuels or industrial greenhouse gases (GHG), manufacturers of vehicles and engines, and facilities that emit 25,000 metric tons or more per year of GHG emissions to submit annual reports to EPA. H.R. 2454, The American Clean Energy and Security Act of 2009 (passed in the House June 26, 2009) would amend the Clean Air Act by establishing a "cap-and-trade" system designed to reduce greenhouse gas emissions (GHG) and would cap emissions from refineries and allow trading of emissions permits ("allowances"). As proposed, H.R. 2454 would require U.S. refiners to purchase emission credits for both their stationary emissions and the subsequent combustion of their fuels (predominantly consumed in the transportation sector). S. 3663, introduced in August 2010, would establish a Natural Gas Vehicle and Infrastructure Development Program to promote natural gas as an alternative transportation fuel in order to reduce domestic oil use.

The prospect of declining motor-fuel demand may persuade operators to idle, consolidate, or permanently close refineries.

INTRODUCTION

The U.S. petroleum refining industry experienced what some have called a "golden age" during the years 2004-2007. During this period, the demand for petroleum products, especially gasoline, increased rapidly both in the United States and world markets. Refiners found favorable price-spreads between heavy and light crude oils as well as between crude oil and refined products. The industry operated plants at nearly maximum capacity and posted record profit levels. Unexpected events such as hurricanes that shut down Gulf Coast refineries, concerns over "peak oil" production, and crude oil price speculation likely contributed to spikes in gasoline prices. During the period, many policy makers expressed the concern that U.S. refining capacity was not increasing rapidly enough to keep up with the expected growth in demand for petroleum products.

U.S. gasoline consumption began declining in 2008, by almost 99 million barrels from the previous year, and another 10 million barrels in 2009.[1] Paradoxically, the United States began importing more gasoline—81 million barrels in 2009.[2] U.S. renewable fuel production (in the form of ethanol) exceeded 256 million barrels, and ethanol imports added nearly 4.6 million barrels.[3]

The concern has now shifted to fears that refining overcapacity may exist in the United States, as the state of, and the outlook for, the petroleum refining industry have changed significantly. Current market conditions have resulted in lower capacity utilization rates and refinery closures. These most recent changes in the conditions facing the industry are consistent with a past performance that has been cyclic. However, mandates for a renewable fuel standard (RFS) and increased corporate average fuel economy (CAFE) could influence permanently reduced refining capacity in coming years.

During an era of increasing crude oil prices and concerns for declining domestic crude oil production, many policy makers advocated energy self-sufficiency. Renewable fuels offered the promise of at least offsetting an increasing demand for transportation fuel. Now, though, the prospect of declining motor-fuel demand may mean that the use of more renewable fuels may influence operators to idle, consolidate, or permanently close refineries.

This report begins by looking at the current production capacity of the refineries operating in the United States, and the sources and changes in crude oil supply. It then examines the changing characteristics of petroleum and petroleum product markets and identifies the effects of these changes on the refining industry, including tax considerations. The report concludes with discussion of the policy and regulatory factors that are likely to affect the structure and performance of the industry during the next decade.

BACKGROUND—REFINERIES AND CAPACITY

After a volatile decade marked by record crude oil prices and profit margins, U.S. refiners now face the prospect of possibly long-term decreased demand for their products. Refiners are responding by cutting costs, reducing capacity utilization, and closing facilities.

A decade ago, 158 refineries operated in the United States and its territories. By the Congressional Research Service's count, the number has declined to 124 refineries that process crude oil into fuels, and in addition, 13 refineries that produce lubricating oils and asphalt.[4] These numbers include three refinery complexes, each made up of two formerly independent refineries joined by pipeline.

Although the number of refineries has decreased, operable refining capacity has increased over the past decade from 16.5 million barrels/day to over 18 million barrels/day. By the Energy Information Administration's (EIA) definition, "operable capacity" includes both operating refineries and idle refineries which shut down temporarily for repair or "turn around" for seasonal adjustment in the product slate (for example, reformulating gasoline from winter to summer blends). In addition, some refinery operators have indefinitely idled their refineries to wait for improving demand.[5]

Petroleum Administration for Defense Districts

During World War II, the War Department (now the Department of Defense) delineated "Petroleum Administration for Defense Districts" (PADD) to facilitate oil allocation. At one time, refineries in each PADD processed crude oil and distributed petroleum products for use in the district. The high rate of merchant-marine tankers lost to Nazi submarines operating along the Eastern seaboard prompted construction of the Virginia and Colonial product pipelines to link the Gulf Coast with the Northeast United States. A network of crude oil and petroleum product pipelines now interlinks the PADDs, making them interdependent.

Crude oil sourcing for U.S. refineries varies over time, but in general, the PADD 1 (East Coast) refineries process crude oil shipped from all over the world. PADD 2 (Midwest) and PADD 4 (Rocky Mountains) increasingly depend on crude oil produced and moved by pipeline from Canada and PADD 3 (Gulf Coast) as well as production from the Rocky Mountain states. PADD 3, the largest refining region, obtains crude oil from the Gulf Coast outer continental shelf, Mexico, Venezuela, and the rest of the world. Permitting issues currently stall a pipeline that would deliver Canadian syncrude (from oil sands) to Gulf Coast Refineries. PADD 5 (West Coast) obtains crude oil primarily from Alaska (by tanker) and California, and through imports. No crude oil pipelines link PADD 1 or PADD 5 with the rest of the country.

Most of the country's gasoline is refined in the Gulf Coast (PADD 3), which makes up nearly 45% of the U.S. refining capacity with 45 refineries processing more than 8 million barrels per day (bbl/d). It is followed by the Midwest (PADD 2) and the West Coast (PADD 5) in refining capacity.[6] The East Coast (PADD 1) has been losing capacity, with gasoline imports meeting a growing portion of demand. **Figure 1** below breaks-out refining capacity by PADD.

A 95,000-mile network of petroleum product pipelines serves most of the United States. This network, separate from the network of crude oil pipelines, distributes refined products to balance the demand and supply conditions in each region. Regional differences in mandated fuel gasoline specifications, however, limit the flexibility of distribution by pipeline. Additionally, PADD 5 is largely isolated from the rest of the United States, especially from the large refineries in PADD 3, resulting in a market that has exhibited higher prices and reduced availability under some market conditions.

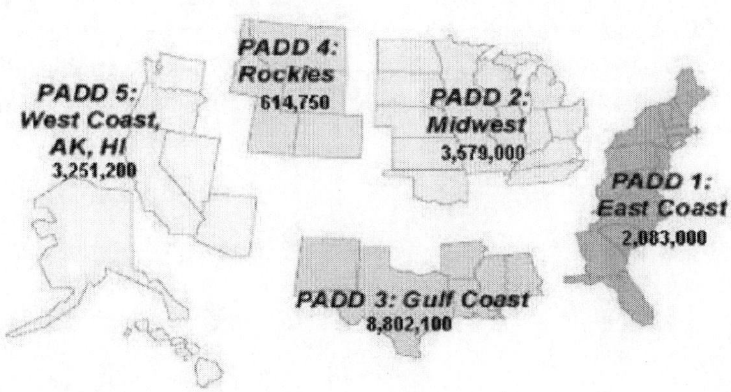

PADD	Fuel Refineries	Bbl/Day
1	12	2,083,000
2	25	3,579,000
3	44	8,802,100
4	15	614,750
5	27	3,251,200
Total	123	18,330,050

Source: CRS.

Note: During World War II, the then-War Department delineated PADDs to facilitate oil allocation.

Figure 1. Fuel Refining Capacity by Petroleum Administration for Defense Districts.

Refinery Closures

After crude oil prices peaked in the summer of 2008, the U.S. demand for refined petroleum products began to decline. In response, U.S. refiners began cutting back capacity and in some cases temporarily idled or permanently closed refineries.

Valero closed its Delaware City (DE) refinery in late 2009 and furloughed 550 workers.[7] In April 2010, Valero sold the refinery to Connecticut-based PBF Energy Partners LLC (Petroplus) for $220 million. Valero will reportedly write off more than $1.7 billion of assets.[8] PBF plans to invest another $125 million to $150 million in refurbishing the refinery with plans to reopen it in the spring of 2011.[9] Sunoco permanently closed its Eagle Point (NJ) refinery in early 2010 and furloughed 400 workers. [10] Sunoco had purchased Eagle Point in 2004 for about $250 million. In 2010, Western Refining idled its 16,800 bbl/day refinery in New Mexico and its 70,000 bbl/day Yorktown (VA) refinery.

Operable Refineries

Figure 2 through Table 5, below, identify operable fuel refineries by PADD.

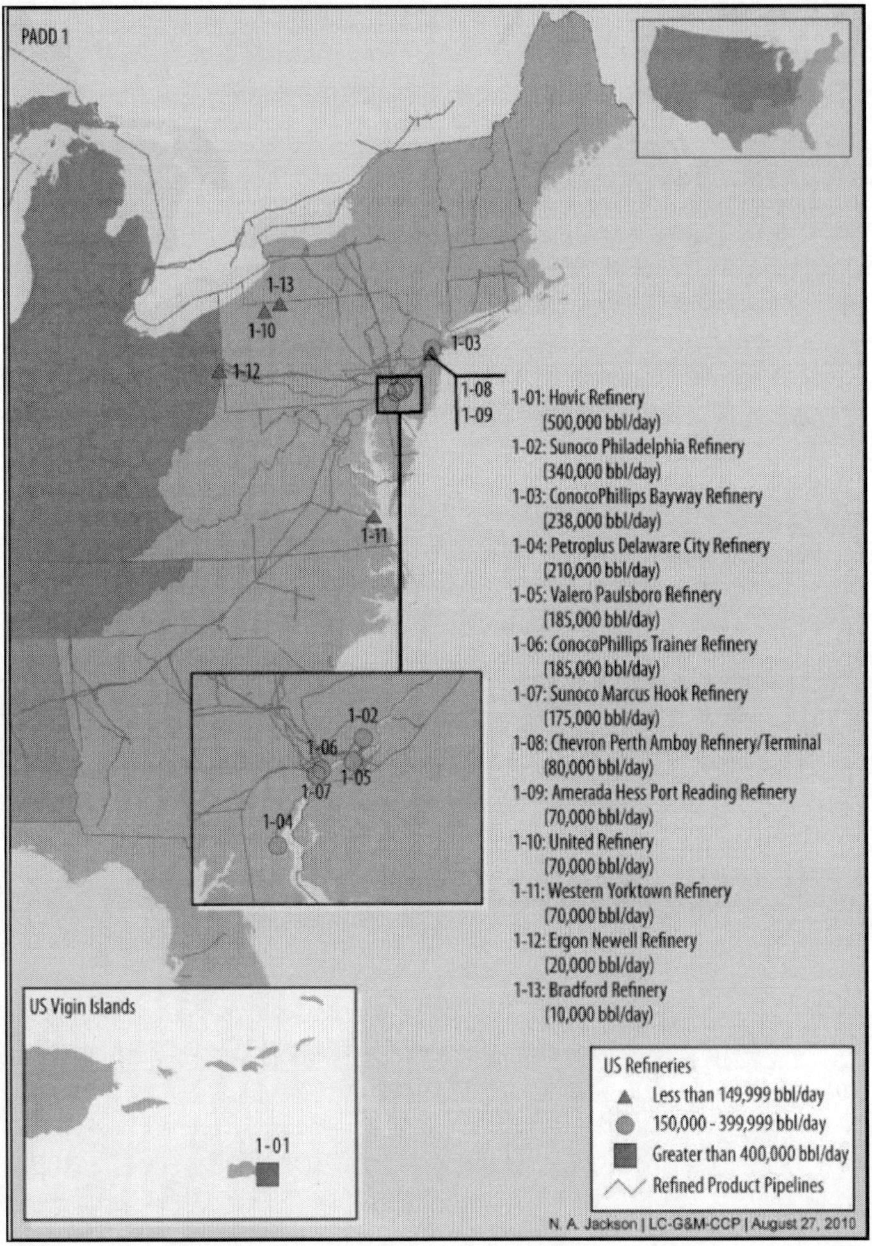

Figure 2. Operable Refineries in PADD 1.

PADD #	Facility	City	State	Zip	Bbl/day
1-01	Hovic Refinery	Kingshill, St Croix	VI	00851	500,000
1-02	Sunoco Philadelphia Refinery	Philadelphia	PA	19145	340,000
1-03	ConocoPhillips Bayway Refinery	Linden	NJ	07036	238,000
1-04	Petroplus Delaware City Refinery*	Delaware City	DE	19270	210,000
1-05	Valero Paulsboro Refinery	Paulsboro	NJ	08066	185,000
1-06	ConocoPhillips Trainer Refinery	Trainer	PA	19061	185,000
1-07	Sunoco Marcus Hook Refinery	Marcus Hook	PA	19601	175,000
1-08	Chevron Perth Amboy Refinery/Terminal	Perth Amboy	NJ	08861	80,000
1-09	Amerada Hess Port Reading Refinery	Port Reading	NJ	07064	70,000
1-10	United Refinery	Warren	PA	16365	70,000
1-11	Western Yorktown Refinery*	Yorktown	VA	23692	70,000
1-12	Ergon Newell Refinery	Newell	WV	26050	20,000
1-13	Bradford Refinery	Bradford	PA	16701	10,000
				Total	2,153,000

Source: Refiner publications.

Notes: The Eagle Point Refinery, which closed in 2010, is not included on this map.

* Petroplus plans to reopen the idled Delaware City refinery in 2011. Western announced August 5, 2010, that it would idle its Yorktown refinery but continue to operate it as a terminal.

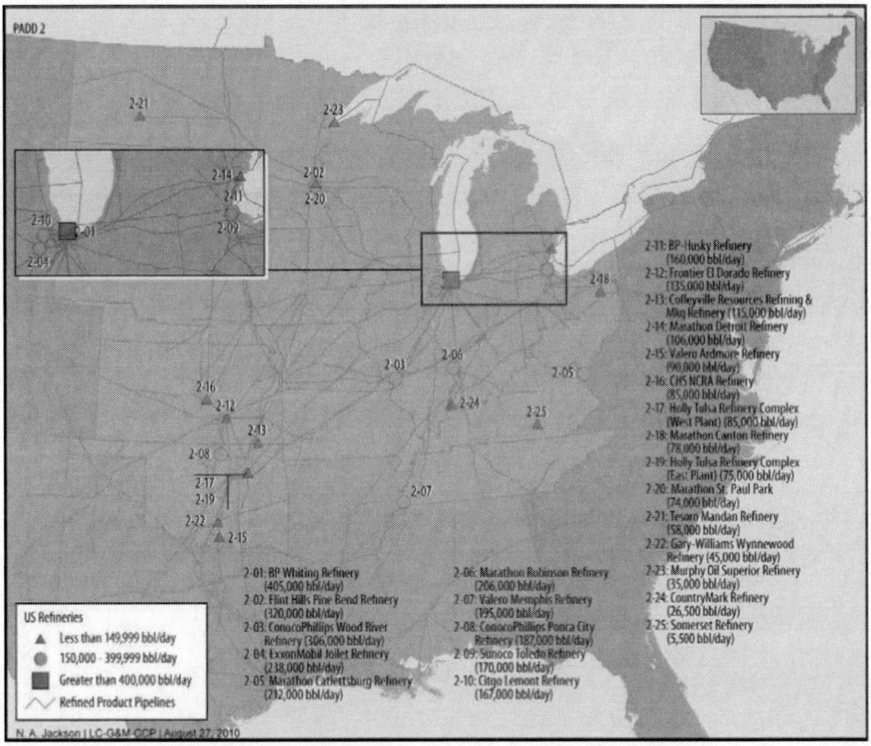

Figure 3. Operable Refineries in PADD 2.

PADD #	Facility	City	State	Zip	Bbl/day
2-01	BP Whiting Refinery	Whiting	IN	46394	405,000
2-02	Flint Hills Pine Bend Refinery	Rosemont	MN	55068	320,000
2-03	ConocoPhillips Wood River Refinery	Roxana	IL	60284	306,000
2-04	ExxonMobil Joliet Refinery	Drummond	IL	60410	238,000
2-05	Marathon Catlettsburg Refinery	Catlettsburg	KY	41129	212,000
2-06	Marathon Robinson Refinery	Robinson	IL	62454	206,000
2-07	Valero Memphis Refinery	Memphis	TN	38109	195,000
2-08	ConocoPhillips Ponca City Refinery	Ponca City	OK	74601	187,000

(Continued)

PADD #	Facility	City	State	Zip	Bbl/day
2-09	Sunoco Toledo Refinery	Toledo	OH	43607	170,000
2-10	Citgo Lemont Refinery	Lemont (Chicago)	IL	60439	167,000
2-11	BP-Husky Refinery	Oregon/Toledo	OH	43616	160,000
2-12	Frontier El Dorado Refinery	El Dorado	KS	67042	135,000
2-13	Coffeyville Resources Refining & Mkg Refinery	Coffeyville	KS	67337	115,000
2-14	Marathon Detroit Refinery	Detroit	MI	48217	106,000
2-15	Valero Ardmore Refinery	Ardmore	OK	73401	90,000
2-16	CHS NCRA Refinery	McPherson	KS	67460	85,000
2-17	Holly Tulsa Refinery Complex (West Plant)	Tulsa	OK	74107	85,000
2-18	Marathon Canton Refinery	Canton	OH	44706	78,000
2-19	Holly Tulsa Refinery Complex (East Plant)	Tulsa	OK	74107	75,000
2-20	Marathon St. Paul Park Refinery	Saint Paul Park	MN	55071	74,000
2-21	Tesoro Mandan Refinery	Mandan	ND	58544	58,000
2-22	Gary-Williams Wynnewood Refinery	Wynnewood	OK	73098	45,000
2-23	Murphy Oil Superior Refinery	Superior	WI	54880	35,000
2-24	CountryMark Refinery	Mount Vernon	IN	47620	26,500
2-25	Somerset Refinery	Somerset	KY	42501	5,500
				Total	3,579,000

Source: Refiner publications.

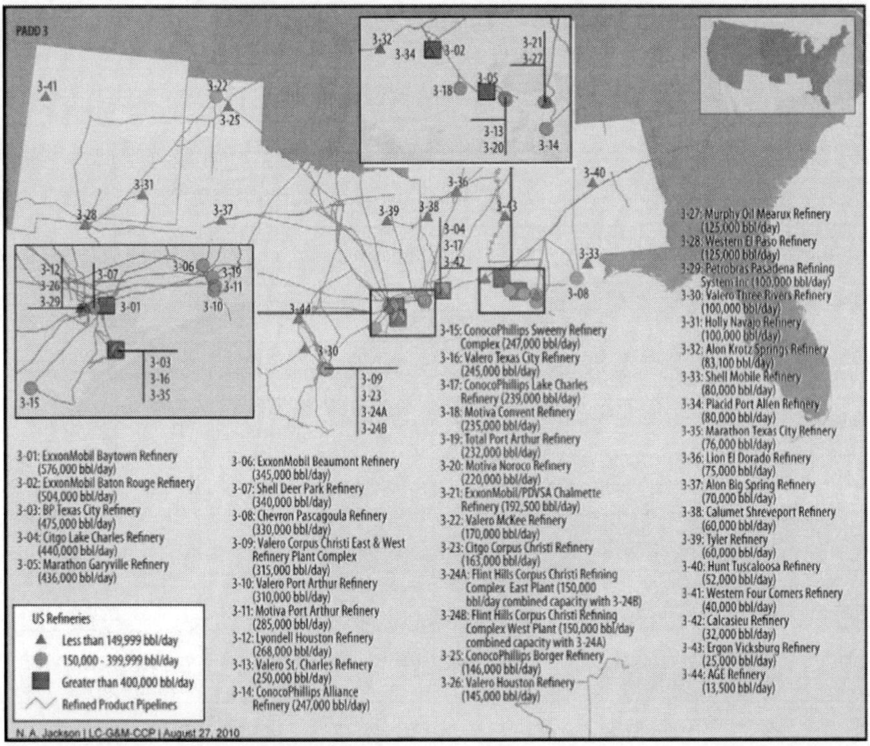

Figure 4. Operable Refineries in PADD 3.

PADD #	Facility	City	State	Zip	Bbl/day
3-01	ExxonMobil Baytown Refinery	Baytown	TX	77520	576,000
3-02	ExxonMobil Baton Rouge Refinery	Baton Rouge	LA	70805	504,000
3-03	BP Texas City Refinery	Texas City	TX	77590	475,000
3-04	Citgo Lake Charles Refinery	Lake Charles	LA	70601	440,000
3-05	Marathon Garyville Refinery	Garyville	LA	70051	436,000
3-06	ExxonMobil Beaumont Refinery	Beaumont	TX	77703	345,000
3-07	Shell Deer Park Refinery	Deer Park	TX	77636	340,000
3-08	Chevron Pascagoula Refinery	Pascagoula	MS	39581	330,000

(Continued)

PADD #	Facility	City	State	Zip	Bbl/day
3-09	Valero Corpus Christi E. & W. Refinery Complex	Corpus Christi	TX	78407	315,000
3-10	Valero Port Arthur Refinery	Port Arthur	TX	77640	310,000
3-11	Motiva Port Arthur Refinery	Port Arthur	TX	77641	285,000
3-12	Lyondell Houston Refinery	Houston	TX	77017	268,000
3-13	Valero St. Charles Refinery	Norco	LA	70079	250,000
3-14	ConocoPhillips Alliance Refinery	Belle Chasse	LA	70037	247,000
3-15	ConocoPhillips Sweeny Refinery Complex	Sweeny	TX	77463	247,000
3-16	Valero Texas City Refinery	Texas City	TX	77590	245,000
3-17	ConocoPhillips Lake Charles Refinery	Westlake	LA	70669	239,000
3-18	Motiva Convent Refinery	Convent	LA	70723	235,000
3-19	Total Port Arthur Refinery	Port Arthur	TX	77642	232,000
3-20	Motiva Norco Refinery	St. Charles Parrish	LA	70079	220,000
3-21	ExxonMobil/PDVSA Chalmette Refinery	Chalmette	LA	70043	192,500
3-22	Valero McKee Refinery	Sunray	TX	79086	170,000
3-23	Citgo Corpus Christi Refinery	Corpus Christi	TX	78047	163,000
3-24A	Flint Hills Corpus Christi Refining Complex E.				
3-24B	Flint Hills Corpus Christi Refining Complex W.	Corpus Christi	TX	78408	150,000
3-26	ConocoPhillips Borger Refinery	Borger	TX	79007	146,000
3-27	Valero Houston Refinery	Houston	TX	77012	145,000
3-28	Murphy Oil Meraux Refinery	Meraux	LA	70075	125,000

PADD #	Facility	City	State	Zip	Bbl/day
3-29	Western El Paso Refinery	El Paso	TX	79905	125,000
3-30	Petrobras Pasadena Refining System Inc	Pasadena	TX	77506	100,000
3-31	Valero Three Rivers Refinery	Three Rivers	TX	78701	100,000
3-32	Holly Navajo Refinery	Artesia	NM	88210	100,000
3-33	Alon Krotz Springs Refinery	Krotz Springs	LA	70750	83,100
3-34	Shell Mobile Refinery	Saraland	AL	36571	80,000
3-35	Placid Port Allen Refinery	Port Allen	LA	70767	80,000
3-36	Marathon Texas City Refinery	Texas City	TX	77590	76,000
3-37	Lion El Dorado Refinery	El Dorado	AR	71730	75,000
3-38	Alon Big Spring Refinery	Big Spring	TX	79720	70,000
3-39	Calumet Shreveport Refinery	Shreveport	LA	71109	60,000
3-40	Tyler Refinery	Tyler	TX	75702	60,000
3-41	Hunt Tuscaloosa Refinery	Tuscaloosa	AL	35401	52,000
3-42	Western Four Corners Refinery	Gallup/Jamesto wn	NM	87347	40,000
3-43	Calcasieu Refinery	Lake Charles	LA	70605	32,000
3-44	Ergon Vicksburg Refinery	Vicksburg	MS	39183	25,000
3-45	AGE Refinery	San Antonio	TX	78205	13,500
				Total	8,802,100

Source: Refiner publications.

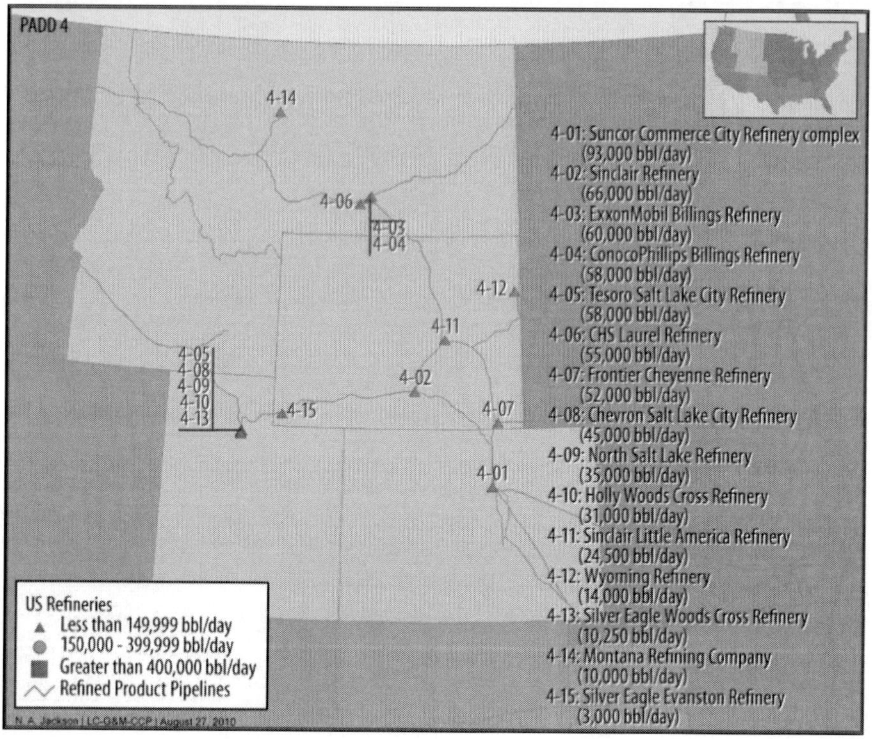

Figure 5. Operable Refineries in PADD 4.

PADD #	Facility	City	State	Zip	Bbl/day
4-01	Suncor Commerce City Refinery Complex	Commerce City	CO	80022	93,000
4-02	Sinclair Refinery	Sinclair	WY	82334	66,000
4-03	ExxonMobil Billings Refinery	Billings	MT	59101	60,000
4-04	ConocoPhillips Billings Refinery	Billings	MT	59101	58,000
4-05	Tesoro Salt Lake City Refinery	Salt Lake City	UT	84103	58,000
4-06	CHS Laurel Refinery	Laurel	MT	59404	55,000
4-07	Frontier Cheyenne Refinery	Cheyenne	WY	82007	52,000
4-08	Chevron Salt Lake City Refinery	Salt Lake City	UT	84116	45,000

PADD #	Facility	City	State	Zip	Bbl/day
4-09	North Salt Lake Refinery	North Salt Lake	UT	84054	35,000
4-10	Holly Woods Cross Refinery	Woods Cross	UT	84087	31,000
4-11	Sinclair Little America Refinery	Casper/Evansville	WY	82609	24,500
4-12	Wyoming Refinery	Newcastle	WY	82701	14,000
4-13	Silver Eagle Woods Cross Refinery	Woods Cross	UT	84087	10,250
4-14	Montana Refining Company	Great Falls	MT	59404	10,000
4-15	Silver Eagle Evanston Refinery	Evanston	WY	82930	3,000
				Total	614,750

Source: Refiner publications.

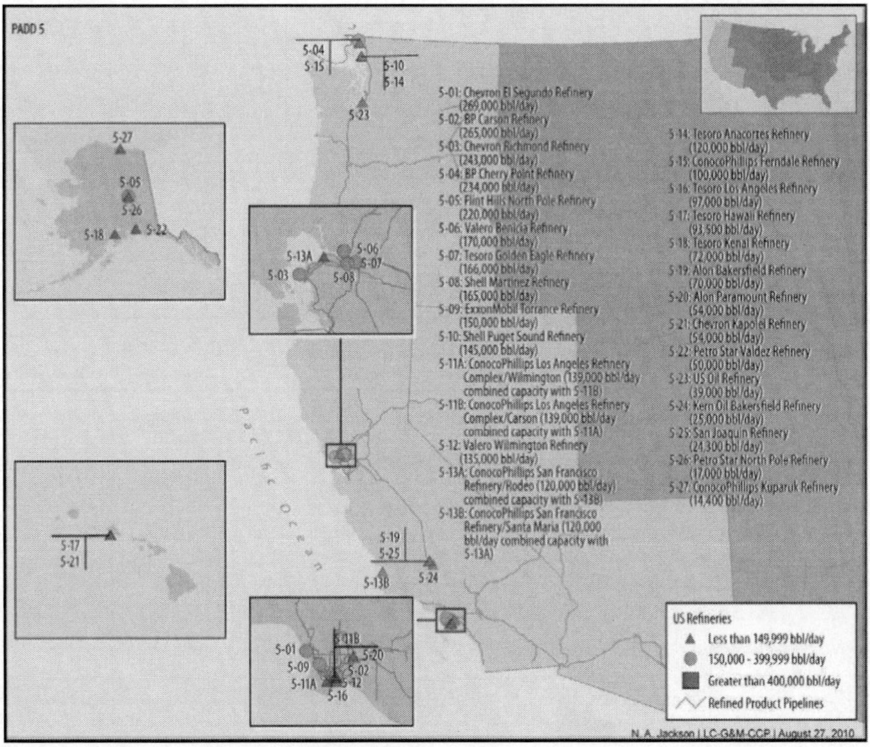

Figure 6. Operable Refineries in PADD 5.

PADD #	Facility	City	State	Zip	Bbl/day
5-01	Chevron El Segundo Refinery	El Segundo	CA	90245	269,000
5-02	BP Carson Refinery	Los Angeles	CA	90745	265,000
5-03	Chevron Richmond Refinery	Richmond	CA	94802	243,000
5-04	BP Cherry Point Refinery	Blaine	WA	98230	234,000
5-05	Flint Hills North Pole Refinery	North Pole	AK	99705	220,000
5-06	Valero Benicia Refinery	Benicia	CA	94510	170,000
5-07	Tesoro Golden Eagle Refinery	Martinez	CA	94553	166,000
5-08	Shell Martinez Refinery	Martinez	CA	94553	165,000
5-09	ExxonMobil Torrance Refinery	Torrance	CA	90509	150,000
5-10	Shell Puget Sound Refinery	Anacortes	WA	98221	145,000
5-11A	ConocoPhillips Los Angeles Refinery Complex/ Wilmington	Wilmington	CA	90744	139,000
5-11B	ConocoPhillips Los Angeles Refinery Complex/ Carson	Carson	CA	90745	
5-12	Valero Wilmington Refinery	Wilmington	CA	90744	135,000
5-13A	ConocoPhillips San Francisco Refinery/Rodeo Facility	Rodeo	CA	94572	120,000
5-13B	ConocoPhillips San Francisco Refinery/Santa Maria Facility	Arroyo Grande	CA	93420	
5-14	Tesoro Anacortes Refinery	Anacortes	WA	98221	120,000
5-15	ConocoPhillips Ferndale Refinery	Ferndale	WA	98248	100,000
5-16	Tesoro Los Angeles Refinery	Wilmington	CA	90744	97,000
5-17	Tesoro Hawaii Refinery	Kapolei	HI	96707	93,500
5-18	Tesoro Kenai Refinery	Kenai	AK	99611	72,000
5-19	Alon Bakersfield Refinery	Bakersfield	CA	93308	70,000

PADD #	Facility	City	State	Zip	Bbl/day
5-20	Alon Paramount Refinery	Paramount	CA	90723	54,000
5-21	Chevron Kapolei Refinery	Kapolei	HI	96707	54,000
5-22	Petro Star Valdez Refinery	Valdez	AK	99686	50,000
5-23	US Oil Refinery	Tacoma	WA	98421	39,000
5-24	Kern Oil Bakersfield Refinery	Bakersfield	CA	93307	25,000
5-25	San Joaquin Refinery	Bakersfield	CA	93308	24,300
5-26	Petro Star North Pole Refinery	North Pole	AK	99705	17,000
5-27	ConocoPhillips Kuparuk Refinery	Kuparuk	AK	99734	14,400
				Total	3,251,200

Source: Refiner publications.

Notes: The ConocoPhillips San Francisco Refinery comprises two facilities inked by a 200-mile pipeline_the Santa Maria facility located in Arroyo Grande, CA, and the Rodeo facility in the San Francisco Bay Area. The Santa Maria facility upgrades heavy crude oil for final processing in the San Francisco Bay facility. The Santa Maria facility is not on the map.

The ConocoPhillips Los Angeles Refinery Complex is composed of two facilities linked by a five-mile pipeline. The Carson facility serves as the front end of the refinery by processing crude oil, and Wilmington serves as the back end by upgrading the products.

Refinery Capacity Distribution

A different picture of the refining industry base emerges when examining the distribution of capacity. As Figure 7 shows, a quarter of U.S. refining capacity is concentrated in a few, larger refineries, reflecting economies of scale that yield decreasing per barrel costs. For example, Royal Dutch Shell PLC plans to double the size of the oil refinery it operates with a Saudi partner in Port Arthur, Texas. This would make it the largest refinery in the United States and one of the largest in the world.[11] ConocoPhillips has plans to expand its Wood River Refinery in Illinois to increase the volume of Canadian heavy crude it can handle, but has run into regulatory hurdles over the use of best available technology under the Clean Air Act. Eleven of the 124 refineries provide one quarter of total U.S. refining capacity. ExxonMobil operates the top two refineries (with a combined capacity exceeding 1 million bbl/day), followed by BP, Petrovesa (PDV), Sunoco, Chevron, Deer Park, and WRB.

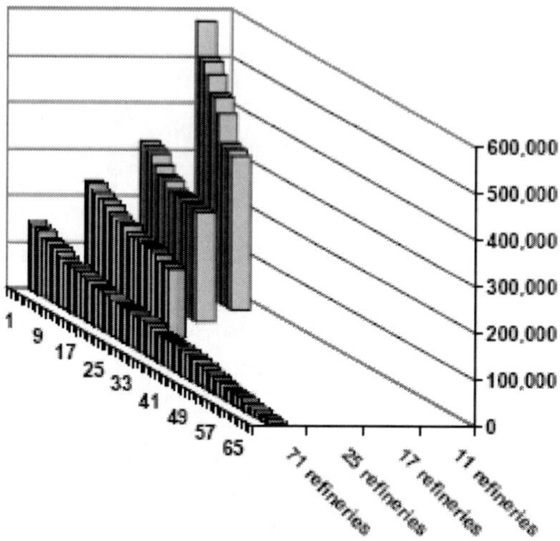

Source: EIA Table 5. Refiners' Total Operable Atmospheric Crude Oil Distillation Capacity as of January 1, 2009, as adjusted by CRS.

Notes: Each quartile represents roughly 4.7 million barrels per calendar day of total refining capacity.

Figure 7. Distribution of U.S. Refinery Capacity; Barrels per Calendar Day.

These eleven refineries are among the largest and most complex in the United States, if not the world, as their owners have added new processes to convert lower value residuum (formerly used as heavy heating oil) to high-value gasoline. Typically, this involves adding fluid or delayed cokers. European refineries, by comparison, employ less complex processes than U.S. refineries on average, as they produce more diesel fuel. (For a further discussion of refinery complexity and processes, refer to Appendix A.)

Changes in Crude Oil Supply and Demand

The crude oil input to U.S. refineries has decreased almost 8% compared to five years ago, reflecting reduced demand for petroleum products. In 2009, refineries consumed an average 14.3 million barrels per day of crude oil. (Refer to Table 1.) Roughly one-third of this input was U.S.-produced in 2009. The balance came from imports supplied by Canada, Saudi Arabia, Mexico, Nigeria, Iraq, and other smaller producers (See Figure 8). Canada has become the United States' leading crude oil supplier through its increasing production from oil sands.[12]

Table 1. Refinery Crude Oil Input (million barrels)

Year	Daily	Annual	Annual Input Change
2004	15.5	5,663.9	
2005	15.2	5,555.3	-108.6
2006	15.2	5,563.4	+8.1
2007	15.1	5,532.1	-32.3
2008	14.7	5,361.3	-170.8
2009	14.3	5,224.3	-137.0

Source: EIA Petroleum Navigator, Petroleum Supply Annual; Refinery and Blender Net Inputs of Crude Oil; http://www.eia.doe.gov/oil psa_volume1_historical.html.

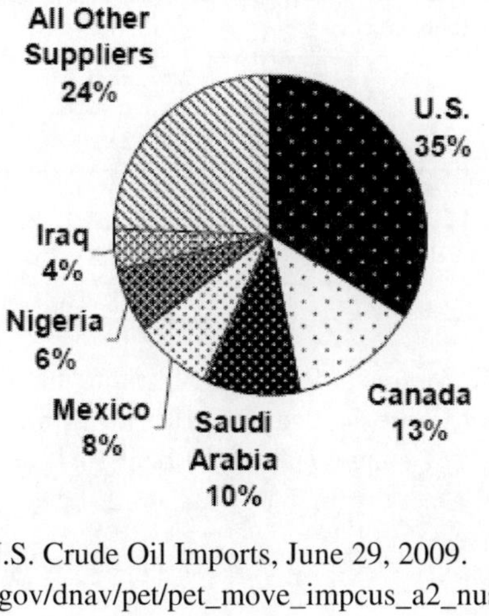

Source: Based on EIA U.S. Crude Oil Imports, June 29, 2009. http://tonto.eia.doe.gov/dnav/pet/pet_move_impcus_a2_nus_epc0_im0_mbbl_a.htm.

Figure 8. U.S. Crude Oil Supply; 2008.

Over the last 25 years, the °API gravity of imported crude oils has been decreasing while average sulfur content has been increasing. °API gravity, a measure developed by the American Petroleum Institute, expresses the "lightness" or "heaviness" of crude oils on an inverted scale.[13] With a diminishing supply of light sweet (low sulfur) crude oil, U.S. refineries have had to invest in multi-million dollar processing-upgrades to convert lower-priced heavier sour crude oils to high-value products such as gasoline, diesel, and jet fuel. Refer to Table 2 for a comparison of various crude oil °API gravities and sulfur contents.

Table 2. °API Gravity and Sulfur Content of Representative Crude Oils

Crude Oil	°API Gravity	%Sulfur
West Texas Intermediate	40	0.30
Alaska North Slope	29.5 – 29	1.10
Strategic Petroleum Reserve sweet/sour	40 – 30	0.5 – 2.0
NYMEX Deliverable Grade Sweet Crude Oil	42 – 37	<0.42
Canadian Sweet/Sour	37.7 – 37.5	0.42 – 0.56
Canadian Alberta Syncrude	38.7	0.19
Saudi Arabia Arab Extra Light / Heavy	37.2 – 27.4	1.15 – 2.8
Mexico Maya/Olmeca	39.8 – 22.2	0.80 – 3.30
Nigeria Bonny Light	33.8	0.30
Iraq Basra Light	34 –35	1.5
Venezuela Tia Juana Light/Heavy	31.8 – 18.2	1.16 – 2.24
North Sea Brent Blend	38 – 39	0.37

Source: NYMEX.

Notes: °API gravity is the American Petroleum Institute's measure of specific gravity of crude oil or condensate in degrees. The measuring scale is calculated as Degrees API = (141.5 / sp.gr.60 deg.F/60 deg. F) - 131.5. Higher API degree indicates lighter, and generally higher priced, crude oils.

Crude Oil Prices

The longstanding benchmark for pricing crude oil futures contracts traded on the New York Mercantile Exchange (NYMEX) has been West Texas Intermediate (WTI) crude oil; a high-quality crude oil with a 39.6° API gravity (making it a "light" crude oil) and a 0.24% sulfur content (making it a "sweet" crude oil). North Sea Brent crude oil, a 38°-39° API gravity light sweet crude oil but with higher sulfur content than WTI, is a global benchmark for other crude oil grades and is widely used to determine crude oil prices in Europe and in other parts of the world.[14] Brent is typically refined in Northwest Europe, and also is exported to the U.S. Gulf and East Coasts.

WTI on average is priced about $1-$2 per barrel above North Sea Brent crude, and $2-$4 per barrel above the Organization of the Petroleum Exporting Countries (OPEC) "basket" of crude prices.[15] OPEC collects price data on a basket of crude oils it produces, and uses the average prices for these oil streams to develop an OPEC reference price for monitoring world oil markets.[16] OPEC's reference basket consists of eleven crude streams representing the main export crudes of all its member countries, weighted according to production and exports to the main markets.[17] According to OPEC, the basket crude has a 32.7° API gravity, making it heavier than WTI or Brent, and a 1.77% sulfur content, making it sourer. Both of these characteristics tend to make it less valuable than WTI or Brent crude. With the diminishing availability of sweet crudes worldwide, U.S. refiners have increasingly turned to heavier sour crudes, and many refineries have upgraded to refine heavier, sourer crudes.

At the beginning of the U.S. invasion of Iraq in March 2003, the spot price for a barrel of WTI crude oil was $28.11, and generally rose during the course of the Iraq War. On a monthly basis, the spot market price of WTI peaked at $133.88 per barrel in June 2008.[18] By February 2009, the price had declined to $39.09 per barrel, only to rise to around $75 per barrel by the end of 2009 and into 2010.

Beside the political uncertainty introduced by the Iraq War, economists have suggested other reasons for the observed price volatility in crude oil markets, including political tensions in Africa and other regions, financial speculation, currency hedging, inflation hedging, excess demand, supply tightness, and a host of other factors. Widely publicized and debated concerns regarding global "peak oil" production may have contributed to speculation in the oil futures market.[19] Because the U.S. dollar serves as the reference price currency for oil in the world market, some oil analysts link the peak in oil prices in mid-2008 to the dollar's weakness at the time. As a result, the oil price rise was much less pronounced when measured in other major currencies.[20]

Although crude oil represents the primary input and cost factor in refinery operations, the relationship between the price of crude oil and the profit margin in refining is neither simple nor direct.[21] Rising crude oil prices increase primary refining costs and can tighten refining profit margins. However, if product prices rise proportionally to crude oil prices, as they did in 2008, refiners effectively pass cost increases on to consumers. Because of the short-term price insensitivity of demand when gasoline prices rise, the revenue derived from the sale of gasoline and other petroleum products is likely to increase in these market conditions, even as total costs are likely to decrease because the volume of oil passing through the refinery declines. These factors can permit refiners to maintain or even increase profits during periods of high crude oil prices. The situation differs if less oil is passing through the refinery due to weak product demand. In that event, product prices and profits may fall in tandem as capacity utilization declines.

The multiplicity of oil prices, which reflect the quality of various crude oils, further complicates the linkage between oil prices and the refining profitability. Generally, lighter crude oils command a price premium over heavier oils, as discussed earlier in this report. The size of the price premium tends to vary as relative supply availability changes and as refiners adapt refineries to use lower cost crude oil stocks. The price spread between light and heavy crude oils, shown in **Table 3**, shrank by almost $10 per barrel between 2006 and 2009.

Table 3. Light/Heavy Crude Oil Price Spread
$ per barrel

Year	Spread
2006	15.51
2007	12.88
2008	14.85
2009	5.60

Source: Energy Information Administration.

Notes: CRS based calculations on North American crude oils, West Texas Intermediate, and Mexican Maya crude.

During the period of high oil prices from 2004 through 2008, heavy crude oils sold at a large discount relative to light crude. The relative tightness in the light crude market, coupled with the price discounts for heavy crude, induced refiners to invest in facilities and processes that would make refineries more able to process heavy crude oils and take advantage of these favorable price spreads. These investments declined in profitability after oil prices fell and the price premium narrowed. In February of 2009, the price-spread declined to a low of $1.93 per barrel, and stayed below $9 per barrel every month in 2009. By September of 2010 the price-spread was $7.95 per barrel.

Demand Conditions

The demand for crude oil is derived from the demand for petroleum products. For example, if consumers demand more gasoline, refiners may purchase and process more crude oil. Afterwards, refiners might adjust their product slate, within technological limits, to yield more gasoline from each barrel of crude oil. (Refer to Appendix A for a discussion of refining fundamentals.)

The demand for gasoline itself may depend upon the price of gasoline and the income level of consumers. However, in the short run, the responsiveness of gasoline demand to variations in price is quite low. Estimates of the short-run price elasticity for gasoline are in the range of -0.25 or less.[22] This value implies that if the price of gasoline rises by 1.0% the result is likely to be only a 1/4 % decline in the quantity of gasoline demanded. Consumers may have difficulty reducing their demand for gasoline in the short-run, as commuting distance, automobile fuel-efficiency, and other commitments make it hard to lower consumption quickly. They may respond to higher gasoline prices by reducing expenditures on other goods or increasing household debt levels. The demand for gasoline also depends on consumer's income growth, and perhaps, as well, on the fraction of consumer's disposable income accounted for by gasoline purchases. The average estimate of income elasticity for gasoline demand in the United States is about 1.0, meaning that a 1% increase in income is associated with a 1% increase in spending on gasoline. Taken together, these elasticity values imply that gasoline demand may increase, even in an environment of high or rising prices, as long as the effect of higher incomes outweighs the effect of higher prices.

This condition appears to have been in place in the United States, and much of the world, during the first half of 2008, as well as much of the 2003-

2008 period in general. However, after the third quarter of 2008 when U.S. gasoline prices had peaked at over $4.00 per gallon, an economic recession coupled with expectations of reduced income growth began moderating the demand for gasoline. After a 0.35% growth in gasoline demand in 2007, demand declined 2.9% the following year, as Table 4 shows.

Table 4. United States Gasoline Consumption 2006-2009 (million barrels per year)

Year	Consumption	Change volume	Change percent
2006	3,377.2		
2007	3,389.3	12.1	0.35%
2008	3,290.1	-99.0	-2.90%
2009	3,280.0	-10.1	-0.30%

Source: Energy Information Administration.

Notes: Gasoline consumption is a measure of product supplied as finished motor gasoline. It includes refinery and blender net production, and imports.

The nearly 3% reduction in gasoline demand, as experienced during the 2007-2008 recession years, may seem minor compared to demand reductions in other industries. Nonetheless, it was sufficient to create the current weak market conditions (characterized by reduced capacity utilization rates, refinery closures, and weak profitability) that the refining industry faces today.

In the longer term, even when income growth returns, the outlook for the gasoline demand in the United States will be constrained by changing attitudes toward petroleum usage, regulations to increase automobile fuel efficiency standards, and regulations mandating the expanded use of alternative fuels in motor transportation.

Profitability

There are 45 firms refining petroleum in the United States. The top 10 refiners—Valero, Conoco Phillips, ExxonMobil, BP, Shell, Marathon, Chevron, Flint Hills, Citgo, and Sunoco—account for more than 75% of total U.S. refining capacity, as shown in Figure 9. The top ten firms operate half of the U.S. fuel refineries, a combined 68 out of 124 refineries (including currently idled refineries).

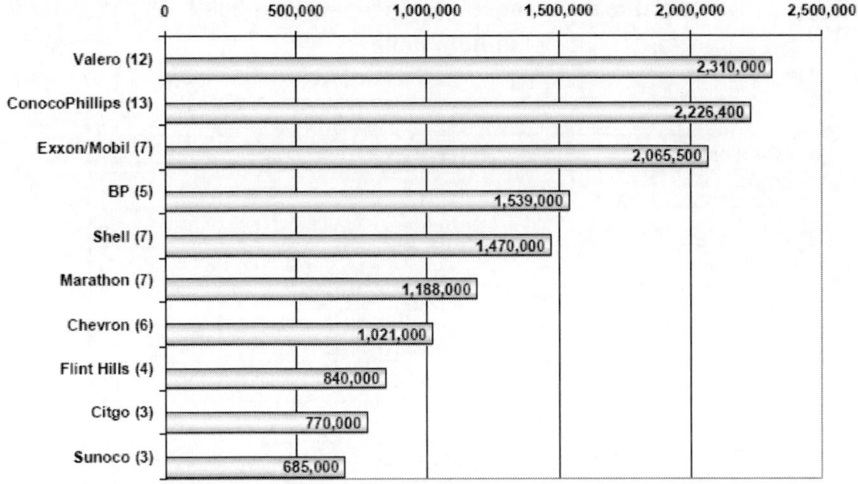

Source: CRS compiled from refiner published information, August 2010.

Notes: Figures in parenthesis indicate number of refineries owned. The top ten refiners represent roughly 75% of the total fuel refining capacity, some 18.5 million barrels per calendar day. This excludes topping and lubricating oils. The privately held Koch Industries owns Flint Hills Resources. The Venezuelan oil company Petrovesa owns Citgo.

Figure 9. Major Refiners by Capacity; Barrels per Calendar Day.

The top six integrated oil companies—ConocoPhillips, ExxonMobil, BP, Shell, Marathon, and Chevron—engage in all phases of the oil business from producing and refining their own oil to transporting it and marketing at retail. Valero, the largest independent refiner and marketer, does not own petroleum reserves. The top six integrated firms plus the top two independent refiners and marketers also make up over 50% of U.S. refining capacity, and control the largest refineries.[23]

Their overall financial performance offers a measure of the profitability in refining and marketing in general. The comparative financial performance for the period 2006 through 2009 is presented in Table 5. The decline in net income over the period is attributable to several factors, including the combination of high crude oil prices and weak demand. The high inventories of gasoline and diesel fuel depressed product prices relative to the cost of crude oil, which further reduced refining profit margins.[24] In addition, the narrowing price spread between light and heavy crude reduced the refining margin and contributed to earlier capital investments failing to generate expected returns.

Table 5. Refiners' Net Income, 2006-2009 (million dollars)

Company	2006	2007	2008	2009
ExxonMobil	8,454	9,573	8,151	1,781
BP	5,667	3,569	4,176	4,517
Shell	6,989	6,624	446	3,054
ConocoPhillips	4,481	5,923	2,322	37
Chevron	3,973	3,502	3,429	565
Marathon	2,795	2,077	1,179	464
Valero	5,461	5,234	-1,131	-1,982
Sunoco	979	891	776	-329

Source: Oil Daily, Profit Profile Supplements, various issues, 2007-2010.

Notes: Data in the table is downstream net income, which includes income derived from refining and marketing. Privately owned Flint Hills and Venezuelan owned Citgo do not publish financial reports.

All six of the major integrated oil companies have experienced mixed returns, but in all cases, their net incomes in 2009 were lower than in 2006. Valero and Sunoco, independent refiners and marketers, experienced losses.

Capital Investment

Refiners undertake capital investment for a variety of reasons, for example, expanding existing or creating new production facilities, implementing new or enhanced technology, or regulatory compliance. Facility expansion and new technology implementation are indicators that the industry expects increasing demand and economic growth.

Capital improvement and expansion require that an initial outlay of funds in the current time-period be offset by earnings that might accrue far into the future. If this stream of appropriately discounted future earning is greater than the initial outlay, then a capital investment project qualifies for inclusion in the capital budget.[25] Because the estimated earnings stream embodies management's forecast of the industry's future economic potential, increasing capital budgets imply expectations of healthy profitability, while declining budgets imply a weak profit outlook.

Capital spending in the U.S. refining sector has been declining, as Table 6 shows. A 22% decline from 2008 through 2009 is expected to be followed by an almost 50% decline from 2009-2010. Combined with refinery closures discussed in this report, this data suggests that the industry does not see a need to expand, or even maintain, production capacity in the United States.

Table 6. U.S. Refining Industry Capital Budget Expenditures, 2008-2010 (billions of dollars)

Year	2005	2006	2007	2008	2009	2010
Expenditure	7.2	9.0	8.3	13.0	10.1	5.3

Source: *Oil and Gas Journal,* Week of March 1, 2010, p. 26.

Refinery Investment and Petroleum Product Imports

While imports of crude oil have been an important part of the U.S. energy supply picture for decades, the importance of petroleum product imports has also been rising. Oil companies can meet the demand for petroleum products, such as gasoline, in three basic ways. They can build new refineries, using either domestic or imported crude oil. This strategy puts refinery investment in competition with the companies' other capital projects, but offers the possibility of relatively large increases in supply.

Alternatively, an oil company can expand the capacity of existing refineries. Investment in expanded capacity can run parallel to investments made to keep existing refinery assets in compliance with environmental and other regulations affecting the industry. Expansions can usually be brought on line faster than new refineries due to simplified permitting requirements, but have the disadvantage of augmenting capacity in smaller steps.

Instead of investing in new refineries or expanding existing ones, an oil company might choose to meet petroleum product demand by importing finished, or partially finished, products from other areas of the world. The advantage of this approach is twofold. The imported products can be introduced, relatively quickly, into the domestic market with no requirement for additional capital spending. The imports can be easily expanded, or contracted, should the need arise. Reliance on foreign sources for petroleum products as well as crude oil adds an additional dimension to concerns of energy dependence, even though prices of these products may be the same in domestic and foreign markets.

Cost is likely to determine an oil company's decision on which alternative to use to meet demand variations. If products available on the world market can meet mandated domestic specifications and are available at competitive prices, importing them gives an oil company flexibility while avoiding the long-term commitment of expanding existing, or constructing new capacity.

A look at U.S. total motor gasoline imports over the 2004-2009 period shows that they averaged about 11% of the roughly 9 million barrels per day finished motor gasoline products supplied to U.S. consumers (see Table 7). Total petroleum products imports made up about 17% of domestic consumption. The effects of the recession can be seen in the reduced level of imports in 2008 and 2009. Adjustments in imports to reflect reduced demand are likely to be accomplished with fewer losses in domestic employment and economic dislocations than refinery closures.

Table 7. Gasoline Imports Vs. Total Gasoline Supplied (Thousand Bbl/Day)

Product	2004	2005	2006	2007	2008	2009
Finished Motor Gasoline Imports	496	603	475	413	302	223
Motor Gasoline Blending Component Imports	451	510	669	753	789	719
Total Gasoline Imports Subtotal	947	1,113	1,126	1,166	1,091	942
Total Finished Motor Gasoline Supplied	9,105	9,159	9,253	9,286	8,989	8,997
Total Petroleum Product Imports	3,057	3,588	3,589	3,437	3,132	2,678

Source: U.S. Energy Information Administration, U.S. Imports by Country of Origin, http://www.eia.gov/dnav/ pet/pet_move_impcus_d_nus_Z00_mbbl_a.htm; and Refiner Motor Gasoline Sales Volumes. http://www.eia.gov/dnav/pet/pet_cons_refmg_d_nus_VTR_mgalpd_a.htm, Product Supplied http://www.eia.gov/dnav/pet/pet_cons_psup_dc_nus_mbblpd_a.htm.

Notes: Other products include fuel oils, pentanes, LPG, unfinished oils, oxygenates, fuel ethanol, kerosene, naphtha, waxes, and lubricants.

TAX CONSIDERATIONS[26]

Provisions adopted in the Energy Policy Act of 2005 (EPAct05; P.L. 109-58) allowed taxpayers to expense 50% of qualified investments in refinery assets.[27] Congress adopted this provision to address concerns that domestic refineries would not have the capacity to meet anticipated growth in domestic fuel demand; a condition that has since reversed. The potential for fuel price spikes also rises when domestic refineries operate at near capacity, as there may be insufficient spare capacity to make up for a refinery outage.

The provisions allowing taxpayers to partially expense investments in refinery assets was initially enacted on a temporary basis.[28] Specifically, taxpayers making qualified investments in domestic refinery property used to refine liquid fuel from crude oil (or other qualified fuels) were eligible for the tax deduction if a binding contract for construction of the qualified property had been entered into by January 1, 2008.[29] Further, under EPAct05, it was required that qualifying property be placed in service prior to January 1, 2012. The Emergency Economic Stabilization Act of 2008 (EESA; P.L. 110-343) extended the under-contract and placed-in-service deadlines, such that the incentive is now available for refineries that entered into a binding construction contract before January 1, 2010, and will be placed in service by January 1, 2014.

Allowing taxpayers to expense part of their investment in refinery property reduces the cost of construction, encouraging additional refinery investment. Allowing 50% of refinery investments to be expensed, rather than depreciated over the normal 10-year life, reduces the cost of construction by approximately 5% for taxpayers in the 35% tax bracket.[30] Since the provision is temporary, there is an incentive to speed up the investment in refinery capacity so as to qualify for the tax incentive. Nevertheless, the incentive to speed up investment is limited, because the effective price discount is small. Investing in excess capacity that would not otherwise be desirable would either leave the plant idle or provide too much output and lower prices and profits for a period of time. The latter cost should be at least as big as the cost of remaining idle.

With a 5% price discount, the interest cost of carrying excess capacity or losing profits could offset the tax credit's value.

The estimated reduction in federal receipts associated with provisions allowing taxpayers to expense 50% of qualified investments in refinery assets is presented in Table 8. Over the five-year 2009 through 2013 budget window, estimates suggest this provision will cost $3.4 billion.[31]

Table 8. Tax Expenditures for Provisions Allowing Partial Expensing of Refinery Investments billions of dollars

	2008	2009	2010	2011	2012	2013
Revenue Loss	0.4	0.5	0.7	0.8	0.7	0.6

Source: Joint Committee on Taxation.

Notes: Tax expenditures are estimate federal revenue losses associated with special tax provisions.

POLICY CONSIDERATIONS

The conventional gasoline refined today has changed considerably since the Clean Air Act of 1970 prohibited lead additives, and later amendments created demand for oxygenated gasoline and reformulated gasoline (RFG). Each of the three formulations of gasoline (conventional, oxygenated and reformulated) is available in at least three grades (87, 89-mid grade, and 91+ super) and the volatility is adjusted for winter/summer and northern/southern driving conditions. (Other properties such as Reid Vapor Pressure, octane, and cetane are discussed in Appendix A.)

Reformulated Gasoline

The Clean Air Act, as amended in 1990, directed the Environmental Protection Agency (EPA) to designate areas not complying with national ambient air quality standards (NAAQS) as ozone "nonattainment areas."[32] Cities with the worst smog pollution are required to reduce harmful emissions that cause ground-level ozone by using reformulated gasoline (known as RFG), which is blended to burn cleaner by reducing smog-forming and toxic pollutants during the summer ozone season. Reformulated gasoline undergoes additional processing to remove volatile components that contribute most to air pollution, and to make it less prone to evaporation. It also contains chemical oxygen, known as oxygenate, to improve combustion.

Since the Clean Air Act Amendments, a growing number of distinct types of gasoline ("boutique fuels") have entered the supply chain. Currently, 15 distinctly formulated boutique fuels are required in portions of 12 states (see Figure 10). In addition to the federal RFG standards, State Implementation

Plans to improve air quality require low-Reid Vapor Pressure conventional gasoline (referred to as "low-RVP"). (Refer to Appendix B for a discussion of RVP and other fuel properties.) California mandates a cleaner fuel than federal RFG (referred to as California RFG, or CaRFG), and the Midwestern states require a unique ethanol-blended RFG.

In analyzing the proliferation of gasoline types, EIA concluded in 2002 that: "... the general impact of an increasing number of distinct gasoline fuels with smaller demands and, in some cases, served by fewer suppliers has been to reduce the flexibility of the supply and distribution system to respond to unexpected supply/demand shifts."[33] The prospect that more refineries may sit idle or permanently close due to decreased demand could further reduce that flexibility.

Source: EPA.

Notes: Currently, 12 states have 15 boutique fuels. Alaska and Hawaii do not have RFG areas.

Figure 10. Map of Reformulated Gasoline Areas.

To reduce the proliferation of boutique fuels, the 2005 Energy Policy Act[34] amended the Clean Air Act (in 42 U.S.C. 7545) by limiting them to the number existing as of September 1, 2004.

H.R. 392, the Boutique Fuel Reduction Act of 2009, would further amend the Clean Air Act Section 211(c)(4)(C)(ii)II) to add temporary waivers for unexpected problems with distribution or delivery equipment necessary for transporting fuel or fuel additives. Amendments to Section 211(c)(4)(C) would give the EPA Administrator authority to reduce the number of boutique fuels after determining that a particular fuel is no longer included in a state implementation plan or is identical to a federally approved fuel.

Between 1992 and 2005, EPA also mandated oxygenated fuel blends to reduce ground-level ozone and smog. Much of the gasoline sold in the United States during that period was blended with up to 10% methyl tertiary-butyl ether (MTBE) as the oxygenate in almost all RFG outside of the Midwest, while ethanol was used in the Midwest. Both MTBE and ethanol served several functions: as an oxygenate in RFG, as an octane booster, and as a volume extender in conventional gasoline.[35] Groundwater contamination concerns and the State of California's ban on MTBE as a gasoline additive left ethanol as the most popular fuel oxygenate. MTBE was produced and added at the refinery. However, ethanol's corrosive nature makes long-distance shipment of ethanol mixed into gasoline impractical. In consequence, ethanol (produced mostly from corn fermentation) is blended with gasoline at the storage terminal where the fuel is dispensed to the fuel tank truck. The shift from MTBE to ethanol thus contributed to a reduction in refinery production.

Renewable Fuel Program/Alternative Fuels

During an era of increasing crude oil prices and concerns for declining domestic crude oil production, many policy makers advocated energy self-sufficiency. Renewable fuels offered the promise of, at least, offsetting an increasing demand for transportation fuel. Now, though, the prospect of declining motor-fuel demand may mean that the use of more renewable fuels may influence operators to idle, consolidate, or permanently close refineries.

Congress created the Renewable Fuel Program under Title XV of the Energy Policy Act of 2005 (EPAct—P.L. 109-58) to substitute increasing volumes of renewable fuel for gasoline. The U.S. Environmental Protection Agency (EPA) has the statutory authority for administering the National Renewable Fuel Standard (RFS) program. The act set a target production volume of 7.5 billion gallons of renewable fuels for calendar year 2012. The 2007 Energy Independence and Security Act (EISA) expanded the program to cover transportation fuels in general, extended the program to calendar year

2022, and increased the target volume to 36 billion gallons renewable fuel annually (857 million barrels annually or 2.3 Million bbl/d) (see Table 9 below).

Table 9. EISA Renewable Fuel Volume Requirement

Year	Cellulosic Biofuel Requirement Billion Gallons	Biomass-based Diesel Requirement Billion Gallons	Advanced Biofuel Requirement Billion Gallons	Total Renewable Fuel Requirement Billion Gallons	Total Renewable Fuel Requirement Million Barrels
2008	n/a	n/a	n/a	9.00	214
2009	n/a	0.50	0.60	11.10	264
2010	0.10	0.65	0.95	12.95	308
2011	0.25	0.80	1.35	13.95	332
2012	0.50	1.00	2.00	15.20	362
2013	1.00	[a]	2.75	16.55	394
2014	1.75	[a]	3.75	18.15	432
2015	3.00	[a]	5.50	20.50	488
2016	4.25	[a]	7.25	22.25	530
2017	5.50	[a]	9.00	24.00	571
2018	7.00	[a]	11.00	26.00	619
2019	8.50	[a]	13.00	28.00	667
2020	10.50	[a]	15.00	30.00	714
2021	13.50	[a]	18.00	33.00	786
2022	16.00	[a]	21.00	36.00	857
2023+	[b]	[b]	[b]	[b]	

Source: EPA Renewable Fuel Standard http://www.epa.gov/otaq/fuels/renewablefuels/

Notes: 1 barrel = 42 gallons.

[a] To be determined by EPA through a future rulemaking, but not less than 1.0 billion gallons.

[b] To be determined by EPA through a future rulemaking.

The 2010 requirement of nearly 13 billion gallons of renewable fuels represents more than 9% of 2009 gasoline consumption.

Under current EPA rules, ethanol is blended up to 10% by volume in retail gasoline (E10) and 85% in E85 fuel for use in flex-fuel vehicles (FFV). On October 13, 2010, the EPA partially granted Growth Energy's waiver request application submitted under section 211(f)(4) of the Clean Air Act.[36] The partial waiver allows the sale of gasoline that contains ethanol up to 15% by volume (E15) for use in 2007 and newer model year vehicles. EPA denied the waiver to use E15 in vehicles older than model year 2000, and is deferring a decision on using E15 in model years 2001 through 2006. The E15 fuel must be sold from a separate pump, as is E85. The new ruling would appear to be at odds with the EPAct 2005 provision that limits the proliferation of boutique fuels.

EPA is finalizing RFS regulations for 2011 with specific annual volumes for cellulosic biofuel, biomass-based diesel, advanced biofuel, and total renewable fuel requirements. Although current ethanol production capacity is more than adequate to meet current blending goals, increased biofuel production faces a number of economic, land use, and policy barriers. The feasibility of expanding current ethanol production by another 1 million bbl/d is linked to the ethanol industry's ability to expand under escalating feedstock prices and economic conditions that discourage capital investment. Congress is also looking toward cellulosic ethanol to meet much of the RFS requirements. However, cellulosic ethanol production has technological barriers to overcome before commercial-scale plants can begin operating.

Subsidies and/or Tax Breaks for Renewable Fuel

Some of the federal subsidies and tax breaks favoring ethanol production were reduced by Title XV of the Food, Conservation and Energy Act of 2008 (P.L. 110-246). The ethanol blender tax credit of $0.51 per gallon (which applies to all ethanol blends, including imports) was reduced to $0.45 per gallon in January 2009 under Section 15331 of the act. The $0.54 per gallon import tariff on ethanol, which effectively offsets the blender tax credit when imported ethanol is blended into gasoline in the United States, is set to expire at the end of 2010 under Section 15333 of the act.

Under the strong motor fuel demand conditions that existed when the Food, Conservation, and Energy Act was passed, ethanol was considered to be a means of extending the volume of refined transportation fuels, particularly after the elimination of MTBE. Since then, ethanol has begun to displace refined fuel, albeit under subsidy. If the renewable fuel volume mandate is met by 2022, and the $0.45 per gallon subsidy were to remain in place, the RFS goal of 857 million barrels could represent a $16.2 billion annual subsidy to displace 564 million barrels of refined gasoline (on an energy equivalent basis).[37]

Carbon Emissions/Greenhouse Gas Rules

In 2007, the United States Supreme Court ruled that EPA has the authority under the Clean Air Act to regulate carbon dioxide (CO_2) emissions from automobiles, and directed EPA to conduct a thorough scientific review.[38] After the ordered review, EPA issued a proposed finding, in April 2009, that greenhouse gases contribute to air pollution that may endanger public health or welfare.[39] Though the finding pertained to automobile emissions, it has wide ranging implications.

In response to the FY2008 Consolidated Appropriations Act (H.R. 2764; P.L. 110-161), EPA issued the Mandatory Reporting of Greenhouse Gases Rule.[40] The rule requires suppliers of fossil fuels or industrial greenhouse gases, manufacturers of vehicles and engines, and facilities that emit 25,000 metric tons or more per year of GHG emissions to submit annual reports to EPA.[41] The rule includes final reporting requirements for 31 of the 42 emission sources listed in the proposal. EPA plans to finalize additional source categories listed in the proposal in 2010. The rule establishes the basis for future legislation and regulations that could cap GHG emissions from refineries as well as other industrial sources.

The American Clean Energy and Security Act of 2009

H.R. 2454—*The American Clean Energy and Security Act of 2009* (passed in the House June 26, 2009) would amend the Clean Air Act by establishing a "cap-and-trade" system designed to reduce greenhouse gas emissions (GHG) and would cap emissions from refineries and allow trading of emissions permits ("allowances"). Over time, H.R. 2454's provisions would reduce the cap to 83% of current emissions, forcing industries to reduce emissions by that amount (cap) or purchase allowances from others who would have reduced emissions more than required or offsets from eligible entities not covered by the cap (trade). The bill would allocate the refining industry only 2% of the total emission allowances for the entire U.S. economy.

Petroleum refineries emit approximately 205 million metric tons of CO_2 annually, which (according to the new EPA rule) represents approximately 3%

of the U.S. GHG emissions. The cost of complying with the new EPA rule could be minimal, but the cost of complying with "cap and trade" provisions of H.R. 2454 or similar legislation could be disruptive to the refining industry according to recent analyses by the consulting firm Wood Mackenzie and the Energy Policy Research Foundation, Inc.[42]

As proposed, H.R. 2454 would require U.S. refiners to purchase emission credits for both their stationary emissions and the subsequent combustion of their fuels (predominantly consumed in the transportation sector). U.S. refiners could face competitive disadvantages with refined petroleum products imported from countries where refinery greenhouse-gas emissions are treated differently. In Wood McKenzie's analysis, U.S. refiners would need to purchase roughly 2,000 million credits in 2015, whereas European Union refiners who export their products (predominantly gasoline) to the United States would only need to purchase 3 million allowances.

Clean Energy and Oil Accountability Act of 2010

S. 3663, introduced in August 2010, would establish a Natural Gas Vehicle and Infrastructure Development Program to promote natural gas as an alternative transportation fuel in order to reduce domestic oil use (see Title XX—Natural Gas Vehicle And Infrastructure Development). The program would also offer incentives to convert or repower conventionally fueled vehicles to operate on compressed natural gas (CNG) or liquefied natural gas (LNG).[43]

Natural gas is abundant in the United States and has already been introduced as a transportation fuel for intra-city buses, principally as means of reducing air emissions. U.S. automobile manufactures marketed passenger vehicles modified to run on compressed natural gas in the 1990s.[44]

U.S refineries currently produce 1,386.5 million barrels of diesel fuel annually. Approximately 5.3 trillion cubic feet (tcf) of natural gas would be needed to replace this fuel, as S. 3663 proposes.[45] In 2008, the United States produced 20.8 tcf of natural gas. To displace its diesel fuel with natural gas, the United States would need to increase natural gas production by more than 25%. This does not take into consideration policies aimed at replacing coal-based electricity generation with natural gas, nor any lost efficiency in converting diesel engines to natural gas.

A refined barrel of 35° API crude can yield about five gallons of diesel fuel, or 12% of the barrel. Displacing all diesel fuel consumption with natural gas represents about two million bbl/day in refining capacity. Refineries cannot cut back diesel production without cutting back on production of gasoline and other refined products. Assuming no decreased gasoline demand, refiners would likely export the excess diesel or market it as heating oil.

Vehicle Efficiency/Mileage Rules

The 2007 Energy Independence and Security Act (EISA) amended the "corporate average fuel efficiency" (CAFE) standards. By 2020, a manufacturer's combined fleet of passenger and non-passenger vehicles must achieve an average 35 mpg. The American Council on an Energy-Efficient Economy estimates that the new standard will save 2.4 million bbl/d by 2030, the equivalent of 13% of the current U.S. refinery output. The EPA and the National Highway Transportation Safety Administration (NHTSA) recently published final rules to implement the first phase of these new standards. Further, they have announced their intention to improve fuel efficiency and reduce greenhouse gas (GHG) emissions for commercial trucks and to adopt the second phase of GHG and fuel economy standards for light-duty vehicles.

Economists have recognized that improving energy efficiency releases an economic reaction that partially offsets the original energy savings, a "rebound effect."[46] According to the National Highway Traffic Safety Administration (NHTSA),

> improving a vehicle's fuel economy reduces its fuel cost per mile driven. In response to the reduced per-mile cost of driving a more fuel-efficient vehicle, some buyers will increase the amount of driving they do, although the precise magnitude of this response is uncertain. Thus imposing stricter fuel economy standards can increase the annual number of miles driven."[47]

Research on the magnitude of the rebound effect in light-duty vehicles dating to the early 1980s concluded that a statistically significant rebound effect occurs when vehicle fuel efficiency improves.[48]

Mathematically, the rebound effect is equal to the elasticity of average vehicle use with respect to fuel cost per mile driven, although the rebound effect is customarily expressed as a positive percentage. NHTSA found that two-thirds of all rebound estimates it reviewed fell in the range of 10% to 30%.[49] NHTSA also cited recent evidence "that the rebound effect has been declining over time, and may decline even further over the immediate future if income rises faster than gasoline prices." In light of the various study results NHTSA reviewed, it elected to use a 10% rebound effect in its analysis of fuel savings and other benefits from higher CAFE standards for MY2012-MY2016 vehicles. The EPA has chosen a more conservative 5% effect.

Will rebound—that is, increased driving—stimulate additional demand for refined petroleum products, or will renewable fuel mandates offset the demand? The legal mandate for increased ethanol consumption further complicates the effort to improve vehicle fuel efficiency. For example, on the basis of energy content it would take roughly 1.39 gallons of E85 to move a vehicle the same distance as one gallon of gasoline.[50]

With EPA's partial Clean Air Act waiver allowing the sale of E15, drivers may see further decreases in vehicles' advertised mile-per-gallon (MPG) ratings. The blend wall increase may further challenge automobile manufacturers to meet the new CAFE standards and possibly erode demand for refined products.

Conclusion

The petroleum refining industry has a long history of cyclical performance. The most recent downturn closely followed a period many identified as the "golden age" of refining. Cycles in the industry have been historically related to movements in the price of oil, which is the primary cost element in refinery operations, and this will likely remain true in the future.

More urgently, the refining industry faces structural challenges from recent government regulations that aim at directly reducing the demand for the industry's output. Higher gas mileage standards for automobiles, increased ethanol content in gasoline blends, and the expansion in the use of pure bio-fuels suggest that even if economic conditions encourage a period of increasing demand for transportation fuels, the need for refined petroleum products will not necessarily increase proportionately. Electric vehicles, if adopted on a large-scale basis, could reduce the demand for liquid transportation fuels of all types.

These policies were intended, in part, to accommodate the growing demand for refined petroleum products. Now, though, the prospect of declining motor-fuel demand means that the use of more renewable fuels could influence operators to idle, consolidate, or permanently close refineries. This possibility may help explain why some refiners do not see a need to expand, or even maintain, production capacity in the United States.

Because of market forces, technological changes, and regulatory pressures on the refining industry, additional refineries are likely to close even as some of the more technologically complex and efficient refineries are likely to expand. If a trend toward even larger refineries emerges, this could lead to concentration in the industry at least on the national level. In the event such adjustments occur, Congress may wish to monitor competitive conditions in oil refining, and in particular the impact of consolidation on the prices and less choice facing consumers.

APPENDIX A. PETROLEUM AND REFINING FUNDAMENTALS

Crude oil contains natural components in the boiling range of gasoline, kerosene/jet fuel and diesel fuel. A typical 35° API crude, oil as shown in Figure A-1, might contain 27% of the hydrocarbons in the range of gasoline and 13% of the hydrocarbons in the range of kerosene and jet fuel. Average crude oils tend to have more paraffin in the gasoline range and more aromatics and asphaltic in the residuum.

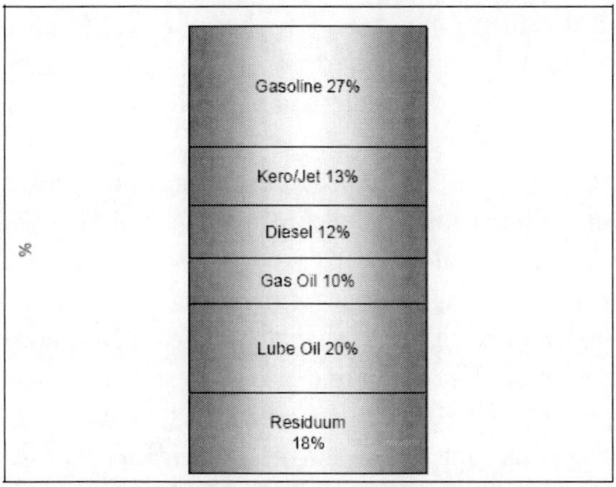

Source: Petroleum Geochemistry and Geology, 1979.

Notes: For illustrative purposes only. Does not represent a specific crude oil assay.

Figure A-1. 35 API Crude Oil Composition.

A conventional refinery distills crude oil into various fractions, according to boiling point range, before further processing. In order of their increasing boiling range and density, the distilled fractions are:

Table A-1. Crude Oil Fractions and Boiling Ranges

Fraction	Boiling Range °F
Residuum	1,050° +
Gas-oil	520° — 1,050°
Kerosene/Jet/ Diesel	380° — 520°
Gasoline /Naphtha	90° — 380°
Fuel Gases	Below 90°

Source: CRS.

Notes: Gasoline's molecular weight is based on the number of carbon atoms, in range of C5 toC10; middle-distillate fuels like kerosene, jet, and diesel range from C11 to C18.

Crude oil may contain 10%-40% gasoline, and early refineries directly distilled a straight-run gasoline (light naphtha) of low-octane rating.[51]

A hypothetical refinery may "crack" a barrel of crude oil into two-thirds gasoline and one-third distillate fuel (kerosene, jet, and diesel), depending on the refinery's configuration, the slate of crude oils refined, and the seasonal product demands of the market.[52]

Conventionally refined gasoline, diesel, and jet fuels are complex mixtures of hydrocarbons that include paraffins, naphthenes, and aromatics (which give fuel its unique odor).[53]

Crude oil processing begins in a refinery's atmospheric distillation unit. The refinery's "name plate capacity," usually expressed as barrels per calendar day or barrels per stream day (see Figure A-2), describes the volume of crude oil that flows through a refinery's atmospheric distillation unit. This is the initial refining stage that separates crude oil into gasoline, kerosene, diesel fuel and heavier petroleum components on the basis of their boiling range. There, the "straight-run" petroleum fractions in the boiling ranges of gasoline, naphtha, kerosene, diesel, and jet fuel condense and separate. Heavier fractions are cracked with catalysts and hydrogen to produce more gasoline range (C5+) blending stock, and low-octane paraffins are converted into high-octane aromatics (octane is discussed below). Other processes such as alkylation produce branched chain hydrocarbons in the gasoline range.

Generally, refineries are set up to run specific grades of crude oil, for example light sweet or heavy sour. Light sweet crude is particularly desirable as a feedstock for gasoline refining because its lighter-weight hydrocarbons make it easier to refine. Heavier crude oils require more complex processing than light crudes, and sour crudes require desulfurization. Refineries upgraded to process heavier crudes cannot readily switch back to lighter oils and run at normal capacity.

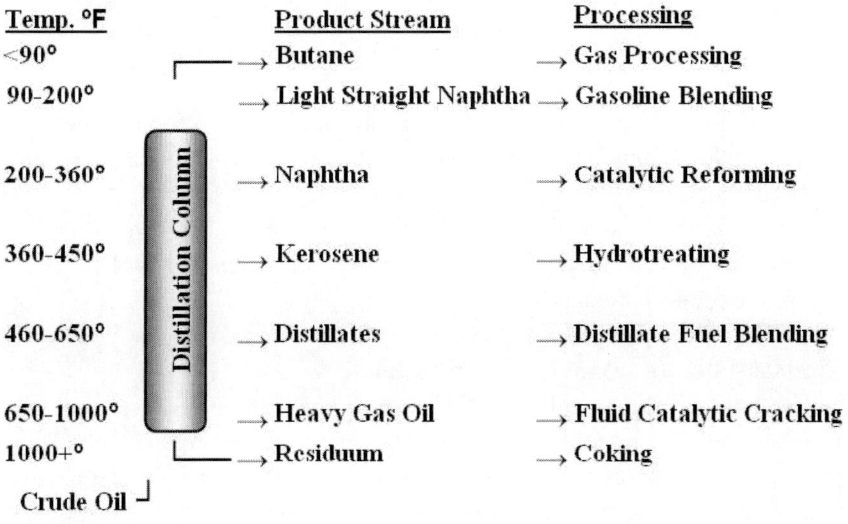

Source: CRS.

Figure A-2. Distillation Column.

Distillation Unit: Heats crude oil until it boils vaporizes. Each hydrocarbon rises to a tray at a temperature just below its own boiling point. There, it cools and turns back to a liquid. The lightest fractions are liquefied petroleum gases (propane and butane) and the petrochemicals used to make plastics and other products. Next come gasoline, kerosene, and diesel fuel. Heavier fractions are used as home heating oil and as fuel in ships and factories. Still heavier fractions are made into lubricants and waxes. The remains, which include asphalt, are known as “residuals.”

Fluid Catalytic Cracker: “Cat cracking” is a refining process used to manufacture gasoline. The process uses intense heat, low pressure, and powdered catalyst to accelerate the chemical reaction of the heavy fractions into smaller gasoline molecules.

Selective Hydrocracker: Partially converts diesel-range material into gasoline, propane and butane via a chemical reaction that uses high temperatures and pressures in a catalyst-containing reactor.

Alkylation Plant: Converts light hydrocarbons to heavier hydrocarbons more compatible as gasoline components for high-octane gasoline.

Catalytic Reforming: A process for upgrading low octane naphtha to a high octane gasoline blending component, reformate. Important by-products of this process include hydrogen, benzene, toluene, and xylenes.

Delayed Coker: Converts petroleum pitch into petroleum coke and gas oils for processing in other units to higher quality, higher value diesel fuel and gasoline.

Gas Oil Hydrotreater: Provides for removal of sulfur and nitrogen from various products, making them more suitable for conversion feed to other process units.

Gas Plants: Collect gases from processing units (hydrocracker, hydrotreater, reformer, coker, cat cracker) and separate volatiles into appropriate product streams.

Sulfur Recovery Unit: Recovers sulfur from refinery streams as elemental sulfur for sale as end-use products.

Catalytic cracking, coking, and other conversion units, referred to as secondary processing units, have enabled refineries to produce more high-value products, such as gasoline, from a barrel of crude oil and process heavier crude oils; see Table A-2. These processing units add to a refinery's complexity and can actually increase the volume of its output. These processes also require a supply of hydrogen, typically derived from natural gas.

A typical refinery yields a limited supply of jet and diesel fuel depending on the type of crude oil processed, see Figure A-3. Gulf Coast (Texas and Louisiana) may yield up to 8% jet fuel, and over 30% diesel. These refineries have an average complexity of 12 to 13, which is above the national average of 9.5.

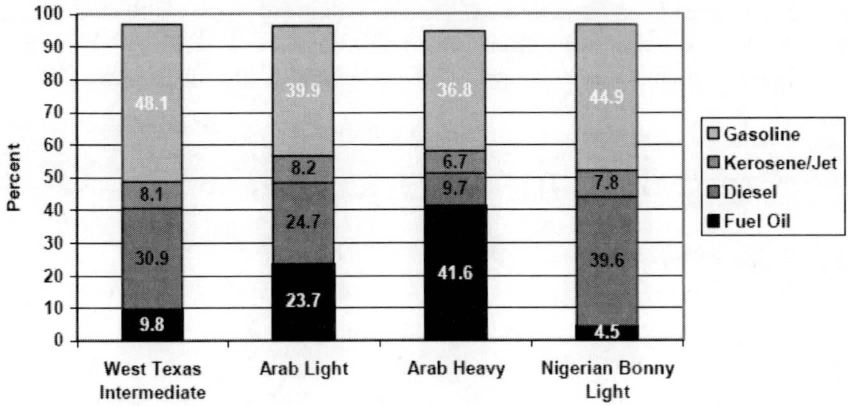

Source: Data used from Energy Intelligence, *The International Crude Oil Refining Handbook*, 2007. http://www.energyintel.com.

Notes: Winter yields shown.

Figure A-3. Gulf Coast Refinery Yields, Percent (%).

Table A-2. Refinery Types and Process

Refinery Type	Processes	Complexity
Coking	Add coking/resid destruction (delayed coking process) to run medium/sour crude oil.	9
Cracking	Add vacuum distillation and catalytic cracking process to run light sour crude to produce light and middle distillates.	5
Hydroskimming	Atmospheric distillation, naphtha reforming and desulfurization process to run light sweet crude and produce gasoline.	2
Topping	Separate crude oil into constituent petroleum products by atmospheric distillation; produce naphtha but no gasoline.	1

Source: Reliance Industries, Ltd., "Types of Refinery & Nelson's Complexity."

Notes: Complexity, as denoted above, is based on the Nelson Complexity Index, which rates the proportion of secondary processes to primary distillation (topping) capacity. Nelson's index varies from about 2 for hydroskimming refineries to about 5 for cracking refineries, and over 9 for coking refineries While the average index for U.S. refineries is 10, only 52 have coking capacity (accounting for the Delaware City refinery closure, this represent 3.485 million barrels per day capacity).54 By and large, U.S. refineries have become the most complex in the world in order to convert low-value residuum, formerly used as heavy heating oil, to high-value gasoline. European refineries, in comparison, are less complex than U.S. refineries on average, being geared toward more producing diesel fuel.

APPENDIX B. IMPORTANT FUEL PROPERTIES

Reid Vapor Pressure

Vapor pressure is an important physical property of both automotive and aviation gasoline, affecting starting, warm-up, and tendency to vapor lock with high operating temperatures or high altitudes. EPA regulates the vapor pressure of gasoline sold at retail stations during the summer ozone season (June 1 to September 15) to reduce evaporative emissions from gasoline that contribute to ground-level ozone and diminish the effects of ozone-related health problems. Shifting to gasoline with lower Reid vapor pressure (RVP) reduces emissions. The Reid Method refers to American Society for Testing and Materials (ASTM) standard test method D 323-08 for measuring the vapor pressure of petroleum products. RVP varies from 8.7 in the summer to 11.5 in the winter.

Octane

Higher octane-number fuels better resist engine "knock"—the sound caused by fuel prematurely igniting during compression. In early gasoline research, the least knock resulted from using isooctane, which arbitrarily received a rating of 100.55 Isooctane refers to a branched "isomer" in the paraffin series having eight carbons (C8H18).56 The straight-chain isomer in this series, n-octane, has a rating -19. Modern formulated gasoline ranges in octane from 87 to 93, achieved by blending various petroleum distillates, reforming gasoline-range hydrocarbons, and adding oxygenates such as ethanol to boost octane-number.

Cetane

The standard for rating diesel fuel's ease of auto-ignition during engine compression is based on "cetane"—a straight-chain hydrocarbon in the paraffin series with the common name of nhexadecane. It consists of 16 carbon atoms with three hydrogen atoms bonded to the two end carbons and two hydrogens bonded to each of the middle carbons; written as C16H34. Pure cetane received the number 100 for rating purposes. Diesel fuel cetane-number ranges from 40 to 45 in the United States to as high as 55 in Europe (where high-speed diesel engines are prevalent in light-duty passenger vehicles). Diesel fuel formulation blends straight-run cut distillates with cracked stock (heavier fractions) to meet standardized specifications developed by the American Society for Testing and Materials (ASTM International) and EPA.

Sulfur

As now regulated by EPA (40 C.F.R. 80.520), diesel fuel must contain less than 15 parts-permillion (ppm) sulfur—referred to as ultra-low-sulfur diesel (ULSD). Conventionally refined aviation jet fuel may contain as high as 3,000 ppm sulfur. However, as it has been used in blending winter diesel fuel to lower the gel point, it has had a practical limit of 500 ppm (the previous EPA limit for diesel). It is uncertain whether EPA may promulgate future rules on jet fuel sulfur-content, thus limiting its use in blending winter ULSD. Despite its detrimental environmental effects, sulfur contributes to the "lubricity" of fuel. Under reduced sulfur, engines wear out sooner. Fuel can be blended with additives to make up for the loss of sulfur lubricity and engines can be manufactured from tougher materials, as has been the case in the EPA mandated transition from low-sulfur diesel (500 ppm) to ultra-low-sulfur diesel (15 ppm). Average annual sulfur content in all gasoline dropped from about 300 ppm in 1997 to about 90 ppm in 2005.

Exhaust Emissions

Diesel engines characteristically emit lower amounts of carbon monoxide (CO) and carbon dioxide (CO_2) than gasoline engines, but they emit higher amounts of nitrogen oxides (NOx) and particulate matter (PM). NOx is the primary cause of ground-level ozone pollution (smog) and presents a greater problem, technically, to reduce in diesel engines than PM. The CO, NOx, and PM emissions for gasoline and diesel engines are regulated by the 1990 Clean Air Act amendments (42 U.S.C. 7401-7671q).

APPENDIX C. GLOSSARY

Motor Gasoline (Finished). A complex mixture of relatively volatile hydrocarbons with or without small quantities of additives, blended to form a fuel suitable for use in spark-ignition engines. Motor gasoline (as defined in ASTM Specification D 4814 or Federal Specification VVG-1690C) has a boiling range of 122° to 158° F at the 10% percent recovery point, and a 365° to 374° F boiling range at the 90% recovery point. "Motor Gasoline" includes conventional gasoline, all types of oxygenated gasoline (including gasohol), and reformulated gasoline, but excludes aviation gasoline. Volumetric data on blending components, such as oxygenates, are not counted in data on finished motor gasoline until the blending components are blended into the gasoline.

Note: E85 is included only in volumetric data on finished motor gasoline production and other components of product supplied.

> **Conventional Gasoline.** Finished motor gasoline not included in the oxygenated or reformulated gasoline categories. Note: This category excludes reformulated gasoline blendstock for oxygenate blending (RBOB) as well as other blendstock.
>
> **Reformulated Gasoline.** Finished gasoline formulated for use in motor vehicles, the composition and properties of which meet the requirements of the reformulated gasoline regulations promulgated by the U.S. Environmental Protection Agency under Section 211(k) of the Clean Air Act. It includes gasoline produced to meet or exceed emissions performance and benzene content standards of federal-program reformulated gasoline even though the gasoline may not meet all of the composition requirements (e.g., oxygen content) of federal-program reformulated gasoline. Note: This category includes Oxygenated Fuels Program Reformulated Gasoline (OPRG). Reformulated gasoline excludes Reformulated Blendstock for Oxygenate Blending (RBOB) and Gasoline Treated as Blendstock (GTAB).

Blendstock for Oxygenate Blending (RBOB). Specially produced reformulated gasoline blendstock intended for blending with oxygenates downstream of the refinery where it was produced. Includes RBOB used to meet requirements of the federal reformulated gasoline program and other blendstock intended for blending with oxygenates to produce finished gasoline that meets or exceeds emissions performance requirements of Federal reformulated gasoline (e.g., California RBOB and Arizona RBOB). Excludes conventional gasoline blendstocks for oxygenate blending (CBOB).

RBOB for Blending with Alcohol. Motor gasoline blending components intended to be blended with an alcohol component (e.g., fuel ethanol) at a terminal or refinery to raise the oxygen content.

Fuel Ethanol (E10). Blends of up to 10% by volume anhydrous ethanol (200 proof) (commonly referred to as the "gasohol waiver").

End Notes

[1] Gasoline consumption, as reported by the U.S. Energy Information Administration, includes blended ethanol.

[2] Reported as 80,882 thousand barrels by EIA.

[3] Reported as 256,149 thousand barrels of fuel ethanol produced and 4,614 thousand barrels imported by Energy Information Administration, *Petroleum Supply Monthly*, February 2010, Table 2 p. 11, http://www.eia.doe.gov/oilpetroleum/datajublications/petroleum_supply_monthly/psm.html

[4] To arrive at this number, CRS used U.S. Energy Information Administration and the Environmental Protection Agency sources, and then cross-correlated information that refinery operators published on their corporate web pages and in financial statements. CRS also geo-located the refinery sites by using online imagery and mapping tools.

[5] The Energy Information Administration defines idle capacity as a component of operable capacity that is not in operation and not under active repair, but capable of being placed in operation within 30 days; and capacity not in operation but under active repair that can be completed within 90 days.

[6] Texas—4,747,179 bbl/day and Louisiana—2,992,123 bbl/day.

[7] Jeff Montgomery, "Valero refinery in Delaware City to close permanently," *The News Journal*, November 20, 2009.

[8] Jeff Montgomery, "Valero announces sale of Delaware City Refinery," *The News Journal*, April 8, 2010.

[9] Steve Goldstein, "Petroplus rallies on deal to buy Delaware refinery," *Wall Street Journal*, April 9, 2010, Market Watch.

[10] "Sunoco idles Eagle Point refinery, furloughs 400 workers, cuts dividend," *The Associated Press*, October 6, 2009.

[11] Texas Gulf Coast Online, *Shell Plans Major Expansion of Texas Gulf Coast Refinery*, http://www.texasgulfcoastonline.com/News/tabid/86/ctl/ArticleView/mid/466/articleld/72/ Default.aspx.

[12] CRS Report RL34258, *North American Oil Sands: History of Development, Prospects for the Future*, by Marc Humphries.

[13] API gravity scale: light - greater than 30°; medium - 22° to 30°; heavy - less than 22°; and extra heavy -below 10°. Formula: (141.5 ÷ relative density of the crude [at 15.5°C or 60°F]) - 131.5.

[14] Commodity Online, http://www.commodityonline.com/commodities/energy

[15] On a daily basis the pricing relationships between these can vary greatly.

[16] PetroStrategies, http://www.petrostrategies.org/Graphs/OPEC_Basket_Crude_Oil_Prices. htm.

[17] The OPEC basket crude oil streams in the basket are: Saharan Blend (Algeria), Minas (Indonesia), Iran Heavy (Islamic Republic of Iran), Basra Light (Iraq), Kuwait Export (Kuwait), Es Sider (Libya), Bonny Light (Nigeria), Qatar Marine (Qatar), Arab Light (Saudi Arabia), Murban (UAE) and BCF 17 (Venezuela).

[18] On a yearly basis, the average price per barrel of WTI rose every year from 2003 through 2008. The daily peak was attained in July 2008, at over $145 per barrel. See WTI Spot Price data at http://www.eia.gov.

[19] For background on the subject of peak oil see Kenneth S. Deffeyes, *Beyond Oil: The View from Hubbert's Peak* (Farrar, Straus and Giroux, 2005).

[20] Steve Hawkes, "Oil nears $100 mark as crude reaches yet another record," *Times Online*, October 30, 2007, http://business.timesonline.co.uk/tol/business/industry_sectors/ natural_ resources/article2767141.ece.

[21] Crude oil generally represents over 50% of the cost of gasoline, the most important refinery product in the United States.

[22] Price elasticity of demand is calculated as the percent change in quantity demanded divided by a specified percentage change in price. The result is a pure number (not measured in any units) that expresses the responsiveness of quantity demanded to changes in the price of the product. A formula to determine price elasticity is e= (percentage change in quantity) / (percentage change in price).

[23] Downstream operations include refining and marketing. Not all petroleum products are marketed by the large oil companies. Some retail outlets are company owned, some privately owned.

[24] Refining margins are the difference between the value of refined products derived from a barrel of crude oil and the cost of refining that barrel. The gross margin subtracts only the cost of crude oil, while the net margin includes all other operational costs as well as crude oil.

[25] This method, which is widely employed by economists and financial analysts, is referred to as *Net Present Value*. An alternative measure is calculation of the internal rate of return to a hurdle rate, usually the company cost of capital.

[26] Molly Sherlock, Analyst in Economics, contributed to this section of the report.

[27] Internal Revenue Code (IRC) § 179C. Under the Modified Accelerated Cost Recovery System (MACRS), petroleum refining assets are depreciated over a 10-year period using a double declining balance method.

[28] For additional background information on energy tax issues, see CRS Report R40999, *Energy Tax Policy: Issues in the 111th Congress*, by Molly F. Sherlock and Donald J. Marples and CRS Report R41227, *Energy Tax Policy: Historical Perspectives on and Current Status of Energy Tax Expenditures*, by Molly F. Sherlock.

[29] Existing refineries may qualify if the installation of new property increases the refinery's capacity by at least 5% or increases the percentage of total throughput attributable to qualified fuels such that it equals or exceeds 25%. All qualifying property must be in compliance with applicable environmental laws on the placed-in-service date.

[30] The present value of a 10-year, double declining balance depreciation per dollar of investment is $0.74 with an 8% nominal discount rate. For every dollar expensed, the benefit of expensing is to increase the present value of deductions by $0.26, and since half of the investment is expensed, the value is $0.13. Multiplying this value by 35% leads to a 4.6% benefit as a share of investment. The value would be larger with a higher discount rate. For example, at a 10% discount rate, the benefit would be 5.4%. The benefit is smaller for firms facing lower tax rates or those with limited tax liability.

[31] U.S. Congress, Joint Committee on Taxation, *Estimates of Federal Tax Expenditures for Fiscal Years 2009-2013*, committee print, 111th Cong., 2nd sess., January 11, 2010, JCS-1-10.

[32] Section 181 of the act required EPA to classify each area as a marginal, moderate, serious, severe or extreme ozone nonattainment area. EPA classified all areas that were designated as in nonattainment for ozone at the time of the enactment of the 1990 Amendments, except for certain "nonclassifiable" areas (56 FR 56694,(1) November 6, 1991).

[33] Energy Information Administration, *Analysis of Selected Transportation Fuel Issues Associated with Proposed Energy Legislation - Summary*, September 2002, http://www.eia.doe.gov/oiaf/servicerpt/fuel/gasoline.html.

[34] Subtitle C—Boutique Fuels Sec. 1541. Reducing the Proliferation of Boutique Fuels.

[35] Environmental Protection Agency, *Status and Impact of State MTBE Bans*, http://www.eia.doe.gov/oiaf/servicerpt/mtbeban/pdf/mtbe.pdf.

[36] EPA, *Partial Grant and Partial Denial of Clean Air Act Waiver Application Submitted by Growth Energy to Increase the Allowable Ethanol Content of Gasoline to 15 Percent; Decision of the Administrator*, October 13, 2010, http://www.epa.gov/otaq/regs/fuels/additive/e15/e15-waiver

[37] Assumes that one barrel of refined crude oil can yield up to 46% gasoline.

[38] *Massachusetts et al. v. Environmental Protection Agency*, 549 U.S. 497 (Supreme Court of the United States, April 2, 2007).

[39] The Proposed Endangerment and Cause or Contribute Findings for Greenhouse Gases under the Clean Air Act was signed on April 17, 2009. On April 24, 2009, the proposed rule was published in the Federal Register (www.regulations) under Docket ID No. EPA-HQ-OAR-2009-0171.

[40] The final rule was published in the Federal Register (www.regulations.) under Docket ID No. EPA-HQ-OAR2008-0508-2278. The rule became effective December 29, 2009.

[41] The gases covered by the proposed rule are carbon dioxide (CO2), methane (CH4), nitrous oxide (N2O), hydrofluorocarbons (HFC), perfluorocarbons (PFC), sulfur hexafluoride (SF6), and other fluorinated gases including nitrogen trifluoride (NF3) and hydrofluorinated ethers (HFE).

[42] Alan Gelder, *The (potentially) Disruptive Impact of Carbon on US Refiners*, Wood Mackenzie, October 27, 2009, http://www.woodmacresearch.com/cgi-bin/wmprod/portal/energy/highlightsDetail.jsp?oid=1611276.

[43] The provision shares objectives similar to the Pickens Plan, which proposes to shift over-the-road trucks and municipal buses from diesel fuel to natural gas. http://www.pickensplan.com.

[44] For further information refer to CRS Report RS22971, *Natural Gas Passenger Vehicles: Availability, Cost, and Performance*, by Brent D. Yacobucci.

[45] The heat content of diesel fuel (139,000 Btu per gallon) refined in 2008 (1,386.5 million barrels) is roughly: (*Eq. 1*) 0.67 × 1,386.5 million bbls × 42 gal./ bbl × 139,000 Btu/gal × therm/100,000 Btu = 5.423 therms. The equivalent volume of natural gas (1,028 Btu per cubic foot) needed to replace the diesel fuel is: (*Eq. 2*) 5,423,239 million Btu ÷ 1,028 Btu/ ft^3 = 5,275.5 million cubic feet.

[46] Kenneth Small and Kurt Van Dender, *The Effect of Improved Fuel Economy on Vehicle Miles Traveled: Estimating the Rebound Effect Using U.S. State Data, 1966-2001*, University of California Energy Institute, Policy and Economics Series, UC Berkeley, CA, September 21, 2005, http://escholarship.org/uc/item/1h6141nj.

[47] National Highway Traffic Safety Administration, *Corporate Average Fuel Economy Compliance and Effects Modeling*, DOT HS 811 112, April 2009, p. 24, http://www.

nhtsa.gov/DOT/NHTSA/Traffic%20Injury%20Control/Articles/Associated% 20Files/ 811112.pdf.

[48] National Highway Traffic Safety Administration, *Corporate Average Fuel Economy for MY2012-MY2016 Passenger Cars and Light Trucks*, August 2009, p. 355, http:// www.nhtsa. gov/DOTAssociated%20Files/MY2012-2016_CAFE_PRIA.pdf.

[49] NHTSA, August 2009, p. 356

[50] E85 has 81,800 BTUs/gal) compared to gasoline's 114,100 BTUs/gal.

[51] Octane number refers to the gasoline property that reduces detrimental knocking in a spark-ignition engine. In early research, iso-octane (C8-length branched hydrocarbon molecules) caused the least knock and was rated 100. Cetane number refers to a similar property for diesel fuel, for which normal hexadecane (C16H34) is the standard molecule.

[52] The term "crack spread" refers to the 3-2-1 ratio of crude-gasoline-distillate. The crack spread and the 3-2-1 crack is a hypothetical calculation used by the New York Mercantile Exchange for trading purposes.

[53] James H. Gary and Glenn E. Handwerk, *Refining Petroleum—Technology and Economics*, 4th Ed., Marcel Dekker, Inc., 2001.

[54] Oil & Gas Journal, 2006 U.S. Refining Survey, December 19, 2005.

[55] John M. Hunt, Petroleum Geochemistry and Geology, W. H. Freeman and Co., 1979. p. 51.

[56] Or more correctly 2,2,4-trimethylpentane.

INDEX

C

D

E

F

G

H

I

J

K

L

M

N

O

P

Q

R

S

V

W

Y